ENERGY LESSONS

Martin Knox

ENERGY LESSONS

A COURSE IN SPECTACULAR
CLIMATE CHANGE SCIENCE
AND TECHNOLOGY
FOR CURIOUS MATURE STUDENTS

Martin Knox

Includes text questions with answers after every chapter

First Published – 2024
This edition published 2024 by Novel Ideas
Brisbane, Qld Australia
www.martinknox.com
Copyright © Martin Knox 2024

The National Library of Australia Cataloguing-in-Publication

Creator: Knox, Martin, author.

Title: ENERGY LESSONS / Martin Knox.

ISBN: 978-1-7636472-0-6 (paperback)

Subjects: Non-fiction
            Energy
            Renewable energy
            Pollution
            Nuclear Energy
            Fossil Fuels
            Government
            Philosophy

All rights reserved

This book is a work of non-fiction.

This book is sold subject to the condition that it shall not, by way of trade or otherwise, be lent, resold, hired out, or otherwise circulated without the publisher's prior consent.
All rights reserved. No part of this publication may be reproduced, stored in or introduced into a retrieval system, or transmitted, in any form, or by any means (electronical, mechanical, photocopying, recording or otherwise) without the prior written permission of the publisher. Any persons, who do any unauthorised act in relation to this publication may be liable to criminal prosecution and civil claims damages. The author asserts his moral rights. Only permissioned users are allowed to copy book pages for educational purposes.

Typeset in Calibri 12pt
Cover artwork Donna Munro Book Design
Original image by Christos Georghiou (Adobe Stock extended licence)
Printed and bound in Australia by Ingram Spark

CONTENTS

Chapter	Title: Energy Lessons	Chapter	Title: Energy Lessons
	Prologue		
1	Peak Oil	22	Nuclear Energy
2	Cold Lake	23	Terrarium
3	Forms of Energy and Converters	24	Solar Panels
4	Recycling Energy and Water	25	Wind Turbines
5	Sharing Energy	26	Air
6	Conserving Water	27	Pollutant Evidence
7	Reducing Energy at Home	28	Ecological Response to Climate Change
8	Using Energy Wisely	29	Living in Space
9	Appetite	30	Refrigeration and Air Conditioning
10	Training for Performance	31	Brisbane Circular Underground Railway
11	Energy in Diets	32	Energy Technology
12	Fast Food	33	The Energy Spectacle
13	Fossil Fuels	34	Climate Change in a Spectacle
14	Energy Conservation	35	Spectacle Subscription Frenzy
15	Bioenergy	36	Is Australia a Nanny State?
16	The Greenhous Theory	37	Why Not a Nanny State?
17	Genesis	38	Climate Images Attract Nanny Investment
18	Big Bang	39	Does Australia Need a Nanny State?
19	Emotional Truth	40	A Spectacle Opposing Energy Minimalism
20	Cause and Effect	41	Saving Your Energy
21	Earth's Climate	42	Trends

DEDICATION

I dedicate this book to my family: Zoe, Tessa, Amani, Uly and Dorian, wanting them to carry forward the flame of energy rationalisation, conservation and curiosity, which I have striven to keep burning. I hope my teaching has illuminated science, philosophy and chemical engineering questions, that my students have brought to my classroom online.

ACKNOWLEDGEMENTS

I am indebted to the following.

Donna Munro has looked after the formatting, cover design and publishing.

Dave Jones caste a critical eye over the manuscript and provided valuable feedback on early drafts.

Brad Ahern discussed with me our understanding of climate science theories.

The chapter on Fast Food is a short story I wrote for Sunnybank Hills Writers' Group.

For the Greenhouse Effect and other climate science topics I referred to my writing in my satirical novel Animal Farm 2, 2021, which introduced a new paradigm of climate science.

I was helped by the philosophy of Guy De Bord, The Society of the Spectacle, 1967. I read this with the University of Queensland's Student Philosophy Association, led by Sam Adams.

My class discussed Al Gore's movie An Inconvenient Truth, 2006 and found beliefs not unlike claims in Genesis in the Bible.

The nutrition data I used in my method for planning diets was kindly provided by Maciej Pomian-Srzednicki.

My writing on Australia's nanny state was partly drawn from my novel 'Turkeys Not Bees', 2022, based on a critique by a Canadian Journalist, Tyler Brule, 2015.

I borrowed the Living in Space chapter from an inquiry in the Science in Society course I developed for the Brisbane School of Distance Education.

I am indebted to the Queensland Department of Education for my experience writing instructional materials under the guidance of experts: Betty Baram (Editor); Bob McAllister (Science); Roger Wooller (Biology); Jan Gentner (Chemistry); Adrian Harding (Manual Arts).

Brisbane's University of the Third Age's group Matters Arising, discussed critically energy policy and climate change technology issues. The discussions were led by Garth Sherman.

AUTHOR BIO

Martin Knox grew up on a farm in Somerset, England. He rode a horse and played rugby. He graduated as a chemical engineer from Birmingham University. His work with energy was in a nuclear power station, in petroleum engineering in Canada, in coal mine development and in transportation. He researched alternative systems of government at Imperial College, London. He became a high school teacher and wrote science textbooks with energy emphasis, published by the Queensland Department of Education.

This book is his tenth book published. He has been writing fiction and satirical novels full-time since 2013: speculative, love, politics, crime, sport, totalitarianism, science and technology. He is involved in public policy-making, has proposed an underground railway for Brisbane, developed ideas for mitigating flooding of the Brisbane River and an anti-memoir of his spiritual enlightenment following Friedrich Nietzsche. He has written about the philosophy of climate science from a phenomenological viewpoint. He discusses current issues at U3A and has studied philosophy with students at the University of Queensland. He attends community development forums.

He blogs ideas from his books and relates them to events in the news. He writes letters, plays the guitar, plays chess and walks in the park by the river where he lives.

He reads classical novels, watches movies and enjoys The Big Bang Theory.

He is divorced with children and grandchildren.

LIST OF NOVELS PUBLISHED

Available from Amazon in Australia, USA, UK and Canada

The Grass is Always Browner (2011)
Love Straddle (2014)
Presumed Dead (2018)
$hort of Love (2019)
Time is Gold (2020)
Animal Farm 2 (2021)
Turkeys not Bees (2022)
Brisbane River Anti-Memoir (2023)
The Camel, The Lion and The Child (2024)
Energy Lessons (2024)

A few passages have been extracted from previous books and included in this book without referencing

PROLOGUE

I worked on oil and coal projects for corporations, until I changed to school teaching. My first posting was at a high school in country Queensland. The Science subject master, Colin Cook, sat in on my first lessons with my classes.

'I want to help you get off on the right foot,' he said. 'I won't interfere. I will sit at the back. After the class I will feedback to you anything I would have done differently.'

'You went fine,' he said afterwards. 'I would have been more definite though. You said 'no-one knows' several times and they aren't used to hearing that from a science teacher.'

'But I couldn't tell them about energy in the future,' I said. 'Nothing is definite.'

'It's their future and they want you to know and tell them what they're in for.'

'There are big changes afoot,' I said. 'I can only tell them the situation now.'

'They need more than that. They need hope that things will turn out alright.'

'I'll do what I can,' I said. 'But our energy future is in turmoil. It's not like forecasting a game of cricket.'

'Stick to the technology. The science, economics and politics are up for grabs.'

As it turned out, I needn't have worried that the future was uncertain. My students were in their final year at high school and had begun to take an interest in the wide world. Strangely, the uncertainty in the world interested them, probably because they had agency and could choose a career to make a difference.

Colin's manner with the students was polite and courteous in front of them, but he invariably was harsh about their poor attitudes to schoolwork, which he attributed to their parents' hostility to schooling.

'They won't remember anything you tell them,' he said. The most you can do for them is something practical that will stick in their minds. You won't get far asking them questions either. Their curiosity has died watching too much television and too many videos. Science to them has to have whiz-bang phenomena. The philosophical side is beyond their interest.'

'The holy grail of science teaching is when students understand the logic of an experiment and the illogic of alternatives,' I said. 'Logic is a skill they can build on.'

My best lessons demonstrated practical experiments at the front bench, or had students experimenting at their benches. Ken, the science department's lab assistant, guarded the science equipment jealously. I would request equipment for my lessons but he wouldn't provide it.

'They won't understand it,' he said, annoying me. I believed that science was easiest to explain in the sequence of famous historical experiments.

He was like the school librarian, who resented lending books to students, having had too many stolen. Sometimes I had to pry materials and equipment out of Ken's control, or help myself when he wasn't there.

After several years of teaching classes face to face, I applied for a position writing online teaching materials for the state's distance education schools. I had been teaching from the State syllabus, a dull document. I looked forward to preparing lessons on my favourite topics. My course was multistrand science with strands of physics, chemistry, biology and earth science. Students who wanted more in depth learning of academic science were advised to enrol in those subjects separately.

The sciences: physics, chemistry, biology and Earth science, had energy as their major unifying theme. Energy has laws, rules and values that are foci for elaboration. Astronomy and science theories such as Big Bang are integrated by energy too. I brought water supply into my topic because, in Australia, energy and water policies often overlap and citizens need a grounding in both.

Distance students engaged with the course materials, teachers and their peers at home, via a video and audio platform. They had little opportunity to sneakily interrupt lessons.

When they studied Einstein's Theory of Special Relativity and discovered their peers in the physics class would not study it because it was too difficult, their self-respect was boosted. They realised that multistrand science had not been dumbed down for them. They found they could understand practical situations without the abstract 'hard science' logic and algebra that were difficult. They wanted to understand climate change and the scientific theories of energy used to rationalize public energy policies.

Final year students in mining were interested in my employment experience as a mining engineer. They were interested in the big machines and high pay of employment in coal mines and oil fields. Coal and oil output was growing and there were plenty of jobs. My best students applied to study science at university.

Students studied online at home. They were a motley crew. A few were from outback properties with the nearest school too far away to attend. Many lived in cities but they were too school-phobic to attend. Some were victims of bullying, or had been excluded because they were bullies, or had nervous behaviour like Tourette's syndrome, or had a chronic medical condition that kept them at home. Some were sports players, dancers, gymnasts, swimmers, musicians, or poets and were coached privately in the mornings. Afternoons they studied distance lessons, enabling them to meet the government's education requirements.

Some students were home-schooled by parents, who checked that my lessons conformed with their religion. They provided additional tutoring by leaders of their faith. There were a few students who were from the Exclusive Brethren sect. They were conspicuous by their diligence, their ability to ask questions and their keen participation in classroom debates.

My course texts were used by several other teachers of multistrand. They were distributed to students by internet, with lessons having demonstrations, expositions, questions and answers and open discussion. I assigned them inquiries which I hoped would bring lifelong learning. After each semester of the two-year course there were exams and the results could contribute to university entrance for those students trying to gain acceptance.

Distance students engaged with the course materials, teachers and their peers at home, via a video and audio platform. They had little opportunity to sneakily interrupt lessons. If a student was disrespectful, or unduly argumentative, I could transfer him or her to a breakout room, where the others couldn't see or hear him or her. Usually this would quieten them down. When I accidentally forgot a troublesome student in a breakout room, when the class was over, I apologised.

I wanted to interest students in my experience of science in the petroleum industry and in coal mining, but environmental opposition was growing and jobs were becoming fewer. There were still opportunities to study energy and water resources at university and I interested students to find employment in the resources industry.

CHAPTER 1 PEAK OIL

Normally I wouldn't meet with a distance student like you face-to-face. After you log in for each scheduled lesson, I speak with you in your class online on your screen. Scheduled classes are networked for cohorts of your age peers, with audio and vision, on the school's platform. Communication is as good or even better than in a school room. I can address you together and individually and see your responses. You can talk to me and to the others on your screen, as well as you could in a classroom.

Our text is this book of printed instruction material, in 41 lessons, including my expositions. Blocks of text are followed by selected responses of model students, extracted from the recorded proceedings, edited into this book for solo reading. The student names are fictional, from previous years of this course.

The purpose of the first lesson on Peak Oil is to gain your interest in energy. Energy supply is often in the news at present and oil is predicted to soon reach a peak.

Here is a true story from my work in petroleum engineering.

'We were drilling an oil well in Canada's Rocky Mountains, with a drill string 6 kilometres long and weighing hundreds of tonnes, turning a bit downhole, with mud circulating out of the hole into the mud pit. When, after about a couple of weeks, the drill stem kicked savagely. We had penetrated a pocket of gas at very high pressure. It was a blowout like a fountain and everyone scattered. The drilling column was propelled out of the hole, accelerating up like an elevator and with a loud boom the drill string was shot high into the sky, like spaghetti. It all landed in a tangle a few hundred metres away.'

Buck raised his hand. The students each displayed a card with their name. I knew of him by reputation as a maverick, the joker in the pack.

'Why did you become a teacher?' he asked.
'I wanted to share my passion for oil and minerals with students. I want you to understand the energy business, for you could soon be an energy user, or possibly a worker in energy.'
'I don't have enough energy to be a worker,' Buck said.
There was laughter.
'Tell us about your passion,' said Sophie, lowering her voice saucily.
I laughed. 'I loved working with energy things, at first. My work was founded on the energy potential of oil. But later I became more interested in people and I took up teaching. Energy interested me and I liked to teach about energy such as fuel consumption in different engines.'
'What is interesting about that?' asked Pamela doubtfully, admiring her fingernails.

I held up a small jar of liquid.
'Oil is powerful,' I told the class. 'Made from oil, a teaspoonful of this petrol has enough energy to power an engine to lift me to the top of Mount Everest. Yes, really. With oil, we can power machines and industries almost without limit. Boy, petrol is cheap. There is a calculation for you to do at the end of this chapter.'
'Before the 'oil shock' in 1973,' I told them, 'all you needed was a hot car and a tankful of gas and you could go anywhere you wanted. You really could!
'But oil suddenly became expensive.
'When the Organisation of Petroleum Exporting Countries, the oil sheiks, figured they could get an increase in price for their oil, the oil importers had to allow them a larger share of the profit. The price increased from about a dollar a barrel, to more than twenty. It shook up world trade and the industry.'

'How could they do that? I'm Kelly.'

'They told their agents to demand an extra twenty dollars a barrel for their oil, when they loaded it on to tankers in the Persian Gulf,' I said.

'Oil was depleting quickly, so the importers paid up,' I said. 'Then a few years later, oil became a cause of pollution, interest grew in slowing down growth and we would only pay less. 'Peak oil' happened because depletion was pulling demand up and pollution was pulling it down. But there was another force against oil growth and this was restraint and conservation. People who recognised the finite social utility of oil began to deplore gas guzzlers and materially extravagant uses of oil. They realised that peak oil was coming, that this would reduce their consumption and by adopting more conservative energy uses they would be prolonging the oil production twilight.'

People deserted energy extravagant technologies in the same way that travellers turned away from hunting safaris, ivory, furs and stuffed animals. Their interest in foreign places became aesthetics and photography rather than illicit collecting of artefacts and samples.

'What does peak oil mean?' asked Gary Leblanc.

'What do you think 'peak oil' means, Gary?'

'I'm not sure if it's the top price or the most production.'

'Most production. I want you now to follow the reading of this chapter in your text. Would you please read. We will follow and stop for discussion from time to time. If you have a question, raise your arm.'

'World oil volume has increased steadily,' read Leblanc. 'An industry forecast is oil demand reaching 116 million barrels per day by 2045. That is the new peak oil and it hasn't happened yet. What do you think has actually happened?

'The quantity of oil produced has increased,' said Kelly.

'Correct, Kelly.'

'Gary, continue reading.'

'The amount of oil remaining is increasing despite depletion. It is a puzzle how that could happen. Demand could turn down any day. But demand has continued to increase. We seem to be standing on a cliff without a way down. What is going to happen?'

'Jessica?'
'Oil is finite, meaning the quantity is limited and will run-out.'
'Correct,' I said.
'Does the price keep going up because trading is between willing buyer and willing seller?' asked Tracey, her voice crisp, biting off the words carefully and precisely.
'Good idea, Tracey,' I said. 'By willing, do you mean they are colluding to increase the price?'
'Not colluding,' she said. 'They want the oil and agree to pay a higher price.'
'That's correct. My explanation of the main players' thinking is probably out of date, so be careful,' I said.

Gary continued reading.

'What follows is not an expert market forecast. I want you to hear about the peak oil and the processes of negotiation that would create the event. I haven't been involved with the oil industry for several years and I am not even sure that peak oil is still being negotiated. I wanted you to know that such a concept has been important in the real world, where you could one day be employed. Thinking about peak oil will help you understand about relative values and how supply and demand is negotiated for commodities and goods traded in markets. The oil price is agreed and then the quantity at that price is agreed.'

'Read on please, Gary.'

'Another type of energy is electricity. Per capita electricity consumption, that is per person, is shown in this table for some of the countries.'

I showed the table on an overhead projector and Leblanc read through the data.

COUNTRY	ELECTRICITY CONSUMPTION PER CAPITA 2021, kWh/year
Iceland	51,304
Norway	24,182
Kuwait	15,294
The Canada	14,546
Sweden	12,515
USA	11,267
Australia	9,143
New Zealand	7,993
Russia	6,864
Germany	6,138
China	5,474
UK	4,266
Turkey	3,350
South Africa	3,216
India	1,025
Nigeria	127

After he had read the table, I asked a question.
'Why do people in some countries use more electricity than in others?'
'Do people in cold countries use a lot of electricity? I'm Biggs.'.
'Yes, certainly. But Australia uses more than you would expect for a warm country,' I said. 'Russia is cold but uses less than Australia. How can that be?'

'Perhaps hot countries use a lot of electricity for air conditioners,' said Leblanc. He was a few years older than the others and insightful.

'Yes, that could be Australia. What about Russia?'

'They can't afford as many heaters or air conditioners as Canada.'

'I agree. The obvious question for the future is how will the less-developed countries, like India, afford to catch up their electricity production?' I said. 'How will countries listed halfway, like the UK, China and Australia, afford to catch up the leaders?'

'Maybe they won't try to catch up,' said Leblanc. It was unlikely, but a good attempt. Students often had difficulty discerning motives of competitors.

'Why wouldn't their people want more electricity?' I asked. 'It's more likely Nigerians can't afford electrical appliances and they don't have a distribution grid.'

'Does the wide gap between energy haves and have-nots threaten World peace?' asked Leblanc.

'It could come between friends, I agree. In Australia, fears of carbon dioxide pollution have blocked export of coal to India.'

'They are probably not happy about it.'

'The poorer countries can't afford to be as pollution-free as countries like Australia would like them to be,' said Jessica. 'It's like our expectations are too high.'

'If we trade with them, they could catch us up,' said Kelly.

'Correct. But the greens want us to stop exporting coal,' I said. 'It's like a betrayal.'

'We can't have the cake of stopping coal and also help poor nations develop,' said Leblanc. It was astute and correctly described our dilemma.

'Thank you Gary. We can't see where the balance lies from one side only,' I said. 'It's difficult to compare OPEC's ability to control oil supply, with the USA's ability to control world demand. Only negotiation finds out where the balance between oil supply and oil demand should be.'

'The oil exporter should be more able to get the price he wants because customers have to fill up their cars,' said Tracey evenly, emphasising her meaning. 'The buyer has to have the oil.'

'That didn't happen with Iraq,' I said. 'The balance was struck when the USA attacked them with operation firestorm. Iraqi oil stayed in the ground for years. The USA got oil from someplace else.'

'The Iraqis didn't have the other exporters with them,' said Kelly, showing a knowledge oil geopolitics. 'The Middle East countries have limited political support to hold out and get higher prices. They need their oil money to buy food and weapons. The USA has diversified its supplies and can refuse to pay an exporter if they ask a price that's too high.'

'Correct. Thank you Jessica. We'll end there today. We have discussed Peak Oil and Per Capita Electricity Consumption. Oil geopolitics and national electricity consumption are complex but you have asked some important questions. The next chapters are more detailed and we will work through the text.

Would you complete Task 1 for tomorrow's lesson and read the next lesson about tar sands.

TASK 1
Calculate energy to power an engine to lift an 80 kilogram person to the top of Mount Everest.

TASK 1 Answer
Gravitational potential energy gained = mgh
= 80 X 32.2 X 8849 = 227,950 kgm/s^2 = 227,950 J
Combustion energy of petrol 45.2 kJ/g
Petrol needed = 227950/(45.2 X 1000) = 5.04g

CHAPTER 2 COLD LAKE

Our first lesson had reviewed political and economic aspects of oil and electricity production.

'We have read that oil is a valuable source of energy,' I said. 'A thimbleful of petrol can lift you to the top of Mt Everest. The reading for today's lesson is about extraction of oil from tar sands, from my experience working in Canada.

The class had increased to 15 students.

'We'll follow the text now and stop for discussion when you have questions.'

'First reader, please?'

'I was a field engineer at my company's pilot plant at Cold Lake, Alberta. We were trialling new technology to extract oil from a huge undeveloped resource, the Athabasca Tar Sands.'

'Is a pilot plant where they grow pilots?' asked Sophie Singleton. *She had not seemed interested so far and her joke reflected her superficial interest. But it was funny.*

'Haha,' I answered. *'By coincidence, the Cold Lake Air Base trained pilots for the Canadian Royal Air Force. The town had 10,000 people and trained about ten fighter pilots per year. There would be more pilots, but the CF-104 Starfighter planes were dangerous to fly and there were accidents.*

'Please continue reading.'

'Cold Lake is 300 kms from Edmonton, in northeast Alberta, beside a large lake. At Cold Lake there was experimental production of oil from a few oil wells, initially and increased to thousands of wells later in a commercial operation. I worked at the oil pilot plant, where we were experimenting to extract heavy oil. At first there were about 50 workers producing a few barrels per day, but it expanded and today the town's population is 15,000 producing 200,000 barrels of crude per day. The oil is difficult to extract, requiring up to 9,000 wells, contrasting with Saudi Arabia producing 100,000 barrels per day from a single well.'

'It shows the high profits the Saudi Arabians could get,' said Gary Leblanc.
'Yes. The importers had held the price down to a few cents per barrel for years. The increase in price enabled the project at Cold Lake to become economical.'
'Next reader, please.'

'The Canadians were developing new production techniques at Cold Lake. An early favourite was 'huff-and-puff', using cyclic steam injection and another was steam line-drive. Another method being talked about while I was there was fire-flooding, which had been successful in other oil fields.'

'How deep is the oil?' asked Biggs.
'Near the outcrop it's down about 100 metres, in unconsolidated sand, dipping and extending along a strike of several hundred kilometres.'
'What's a strike?' Norman asked.
'Strike is where the beds would surface, intersecting with a horizontal plane.'
'What did the oilfield look like?' asked Leblanc. It was good to hear him question the appearance of strike, a complicated concept, difficult to explain in words.
'There's a photo. It shows a row of 'nodding donkey' wellhead pumps that lift the oil to the surface. Later on there were more wells side by side, several hundred hectares of them.
'Read on,' I said.

'The pilot plant area was cleared from the surrounding muskeg swamp. There was a boiler with insulated pipes taking steam to each of the oilwells, where the oil was lifted up by pumps into pipelines that gathered it for treatment and trucking away.'

Wellhead pumps and steam pipelines at the pilot plant. (Source: https://www.arcenergyinstitute.com/wp-content/uploads/Rig-Picture.jpg).

'What did you do there, Sir?' asked Jessica.

'Engineers mostly do thinking,' said Sophie, laughing as if it was a joke.

'Yes,' I said. 'We think about all sorts of things. As an engineer my work was to invent, design, analyse and test equipment. For example, I had an idea to increase efficiency of the pumps by cooling them with water. When it was tested, it worked and was adopted as standard practice.'

'Did the oil flow into the oilwell?' asked Leblanc.

'No. It was lemonade,' said Sophie rudely.

'It did after we heated it with steam,' I said, ignoring her. 'We injected it at the wells.'

'How deep did the steam go?'

'We experimented to find how far down the tubing to perforate to allow steam injection into the oil formation. Hot oil flowed out through the perforations.'

'How did you perforate the tubing?'

'We lowered a machine gun and shot holes through the casing at depths where the oil was.'

'Are you joking?' asked Buck. 'Far out. Did the oil hold out for long?'

'Haha.'

'Did the injection and production cycles last long?' asked Tracey. Her technical understanding impressed me.

'We tried injecting for up to a week and then producing for the same time.'

'What were the working conditions like?' Tracey asked.

'In winter, workers were exposed to very cold air while handling very hot pipes, pumps and equipment. Hot tar coated everything and when it cooled became hard to remove. When there was wind, it could be too cold to work. It was quite dangerous, because to frack the sandstone the steam had to be at very high pressure. The workers were reluctant to go near the wells, fearing a massive eruption if a fracture reached the surface.'

'Why did the workers accept the risks?' asked Jessica, genuinely puzzled.

'The oil fetched maybe AUD50 per barrel and at 100,000 barrels per day the field would earn AUD 5 millions per day.

'The costs would be high, but someone could be getting a lot of money,' said Leblanc. 'Were the workers paid well?'

'They got more than I did, as a rookie engineer with a degree,' I said. 'I wasn't there for the money. It was exciting to engineer a new process.'

'Shooting up oil kills me,' said Buck. 'I'll bet a lot of oil escaped. It could run and hide underground.'

'Here is a calculation for you to compare oil from tar sands at the oil field, in Australia after refining it into petrol, with price at the pump.'

'Thank you Tracey.'

'In this lesson you have learned about oil recovery at Cold Lake. This will help you understand oil production technology and working conditions.'

TASK 2

Calculate the $50 per barrel cost of tar at the oilfield compared with the current pump price of 226.9 c/L for petrol in Australia.

TASK 2 Answer

I US Bbl = 159 litres but the tar can be made into only 50 litres of petrol per barrel.

Because only 50 litres of petrol are obtained from one barrel of tar, petrol would earn 50 x 226.9 = $113.45 per barrel of tar.

The tar would cost $50 per barrel with $63.45 going for refining, taxes and profit.

CHAPTER 3: FORMS OF ENERGY AND CONVERTERS

You have studied energy previously in physics, as the ability to do work, existing in forms which can be converted. This chapter will refresh your memories of the different forms of energy in certain things. You need to match the forms of energy usually in objects. Download the table below and complete it. Then check your answers at the end of the chapter.

TASK 3A

No.	OBJECT	FORM OF ENERGY
1.	Electric current	Electrical
2.	A hot oven	Heat
3.	Flashlight beam	
4.	Jack-in-the-box spring	
5.	Water in a dam	
6.	Lump of coal	
7.	A kicked football	
8.	Uranium fuel	
9.	Kitchen mixer blades	
10.	Petrol	
11.	Molten steel	
12.	Stretched archery bow	
13.	Car parked on a hill	
14.	Hammer hitting a nail	
15.	Moving electrons	

16.	Cinema picture	
17.	Speeding bullet	
18.	Box of matches	
19.	Warm soup	
20.	Cold soup	

Answers to TASK 3A are below TASK 3B.

TASK 3B

Now consider devices that convert one form of energy to another. Some answers are supplied. Complete the table showing inputs to conversion devices and their outputs.

DEVICE	CONVERTS FROM	CONVERTS TO
firework	Chemical energy	Light, heat and sound energy
television	Electrical energy	
match		Heat and light energy
light bulb		
catapult		Kinetic energy
falling bucket	Gravitational potential energy	
electric fire		
human body		
microphone		
atomic bomb		
car engine		

The skill here is to recognise forms of energy going to and from conversion devices.
Complete the table above for our next lesson.
See answers below.

TASK 3A Answers

No	OBJECT	FORM OF ENERGY
1.	Electric current	electrical
2.	A hot oven	heat
3.	Flashlight beam	light
4.	Jack-in-the-box spring	elastic
5.	Water in a dam	gravitational potential
6.	Lump of coal	chemical
7.	A kicked football	kinetic
8.	Uranium fuel	nuclear
9.	Kitchen mixer blades	kinetic
10.	Petrol	chemical
11.	Molten steel	heat
12.	Stretched archery bow	elastic
13.	Car parked on a hill	gravitational potential
14.	Hammer hitting a nail	kinetic
15.	Moving electrons	electrical
16.	Cinema picture	light
17.	Speeding bullet	kinetic
18.	Box of matches	chemical
19.	Warm soup	heat
20.	Cold soup	chemical

TASK 3B Answers

DEVICE	CONVERTS FROM	CONVERTS TO
firework	Chemical energy	Light, heat and sound energy
television	Electrical energy	Light and sound energy
match	Chemical energy	Heat and light energy
light bulb	Electrical energy	light and heat energies
catapult	Elastic energy	Kinetic energy
falling bucket	Gravitational potential energy	Kinetic energy
electric fire	electrical	Heat and light
human body	chemical	Kinetic and heat
microphone	sound	Electrical
atomic bomb	nuclear	Kinetic, light and sound
car engine	chemical	Kinetic, heat and sound

CHAPTER 4: RECYCLING ENERGY AND WATER

'In this chapter you will study a large energy conversion device, a pumped storage dam. You are required to be able to explain how it works using the correct scientific terms and concepts.'

'You have seen how water in a dam has gravitational potential energy. It can be converted to electricity in a turbogenerator. Here is a large pumped storage device, being considered for electricity supply. We will discuss how this works.

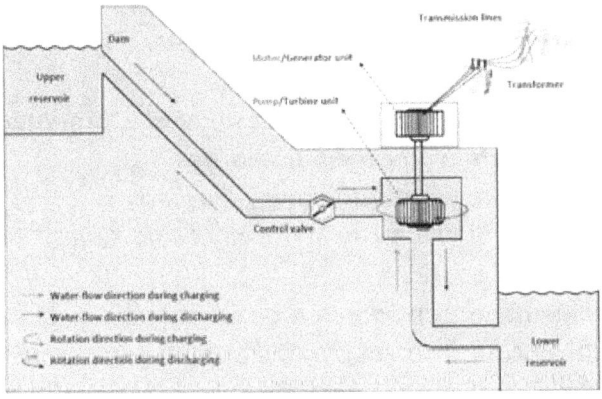

Source ScienceDirect.com (Storage Hydropower)

'Are we doing sports drinks?' asked Buck. 'How much of them is energy and how much is water?'

'We will find out about foods later,' I said. 'First we're going to look at electricity.'

'Battery electricity or mains?' asked Buck.

'Here's a diagram of the device: it's like a giant battery,' I said. 'It's a pumped storage dam generating alternating current.'

I projected the diagram onto their screens.

'It's large. The water could be several hundred metres up.'
I read aloud the labels on the diagram.

'How does it generate electricity?' I asked them.
'Is it like an inflatable swimming pool?' asked Buck, being his obstructive self.
'No. Water is pumped, not air. The grey part of the dam is concrete.'
'Why is water from the bottom pumped up into the upper reservoir?' asked Kelly.
'Why do you think?' I asked, thankful for his question.
'So electricity can run downhill to the city,' said Buck.
'No. Electricity doesn't flow downhill,' I said. 'It's so the water can run down and spin the turbine making electricity.'
'I don't get how water could run in opposite directions through the same pipe,' said Norman slowly.
'At two different times, you nerk,' said Kelly, argumentatively. 'It charges the dam with water then discharges.'
'When would the water run down?' I asked.
'When they want electricity in the city,' said Kelly.
'When is that?' I asked.
'At peak demand time, about 6 pm every day,' Leblanc said. 'The water would run out in a couple of hours. I don't know when they would refill it?'
'They would wait until the next morning and pump the water back into the dam, at say midday, when there was plenty of electricity from solar panels to run the pumps. The dam recirculates water. It can recirculate it many times, converting surplus electricity for peak consumption.'
'Would some water come from rainfall, without being pumped up there,' said Emily. She spoke quietly but I was impressed by her comprehension of technology.

'Correct. Regular dams hold water for years and may generate some hydro electricity but a pumped storage dam can supply more than a hydro dam, emptying many times.'

'Why do we have to know about this?' asked Buck, yawning.

I ignored his rudeness.

'It is a part of our civilisation and because you are curious and may want to have a say in things. You need to know what they can do. There is one of these on the Brisbane River already and there are proposals to build another.

'For our next class, I want you all to write a page about where pumped storage would be most useful and who would benefit. For those with internet access, you could find out about a pumped storage project, or about one being considered.'

The bell for the end of the lesson rang and students packed up.

In the next period we continued the pumped storage chapter, when I projected the diagram again to start the lesson. Some of the students had found out about pumped storage projects.

'Who would like to tell us conditions that would bring benefits from a pumped storage dam?'

'Where both electricity and water are wanted,' a student said.

'That would be dangerous, Sir,' said Sophie. 'Electricity shouldn't be used near water.'

She had been quiet in the class so far and I wondered if she was pulling my leg.

'That's correct, Sophie. We will assume the electricity can be kept away from the water. What could be a situation where a pumped storage dam could be constructed?'

There were no takers for this question.

'Could it save energy? Could it save water?' I prompted.

'One could be constructed to circulate water in a dam through a generator using saved water to save energy.'

'Someone to read, please?'

Energy in the stored water could be converted in the hydroelectric generator to electricity. It could be transmitted to customers in overhead cables. About 90% of the energy in the dammed water would be transmitted to electricity consumers.

'Now you tell me, I interposed, is that efficient? In other words is it reasonable?

The students looked at each other. They hadn't come across efficiency before.

'I'll give you a hint. Where did the water in the dam come from?'

'The water was pumped up,' a student said. 'Could the hydro get back most of the electricity used to pump the water up?'

'Some, not most. Well tried. Would they be able to get something for nothing?'

'No. Energy can't be created,' said Tracey.

'Exactly right,' I said. 'That would be perpetual motion. Perpetual motion is a fantasy. You can't build pumped storage dams, taking electricity away to use in the city, setting them pumping water around in circles. What if the hydro was 90% efficient?'

'They could get back 90% of the energy from the dammed water,' said Leblanc.

'Don't swear, please. I am easily offended,' said Tracey

'I wasn't swearing. I was referring to the dammed water in the dam.'

'Why should you care about it?' said Tracey.

'The electricity could be more useful later,' said Leblanc. 'If the pumping can use cheap abundant electricity from solar panels to fill the dam at midday, in the evening they can run the water in the dam through the hydro, to supply peak demand.'

There was silence, until someone whistled.

Buck said: 'Ooo! I wish I had said that.'

'A pumped storage would not save electricity but it could save money,' I said. 'Could it save water too?'

'It could operate like a hydro, making electricity from the natural flow of the river,' said Emily. Her grasp of the system was excellent.

'Read the next section, please.'

'A pumped storage could use the natural river flow,' I said. 'If water was required for city supply, the pumped storage could add a significant volume. But if the storage was wanted to mitigate flooding downriver, it would need to release the water. Saving water to generate electricity might not be possible for longer than a few hours.'

'I don't get it,' said Norman. *'How could a dam mitigate flooding?'*
'Duh! The dam would hold back the water.'
'Next reader please.'

'It's a complex technology. As we have seen, a pumped storage could be useful. For our next class, would you draw a cross section of a pumped storage hydro, labelling types of energy flowing when the dam is pumped full. The types of energy are gravitational potential, kinetic and electrical. You have seen examples of these in Task 3A but may need to look these terms up and obtain definitions.'
The class ended.
'Bye sir,' said Tracey. *'Thank you. You really care about that dammed water, don't you?'*
'A pumped storage is so cool; I care about it a lot.'
'I would say it's gnarly, not cool at all. It's artificial.'
'You're right. That's technology for you.'

TASK 4
Draw a cross section of a pumped storage hydro, labelling types of energy flowing.

CHAPTER 5 SHARING ENERGY

I asked the class to show me their diagrams of a pumped storage dam and projected the diagram on their screens with the types of energy written on it. There was gravitational potential energy in the dam's water, being converted into kinetic energy in the hydro generator and then into electrical energy flowing in the transmission lines. Most of the students had correct diagrams.

'Well done,' I said. 'I asked you to read the next chapter, about sharing energy with others.
'Next reader please.'

How to share resources such as oil with other communities and nations is resolved by politicians through legislation and in case of dispute, sometimes with military conflict.
'I care a lot about responsible use of energy. In my view, protecting people from running out of energy should be one of the highest priorities of government.
'I want people to realise that energy consumption has adverse consequences for others and they should moderate their consumption responsibly.'

'There's not much responsibility in our community,' Leblanc said. *'At present, in a drought or a power outage, it's everyone for himself.'*
He said it like it was the only possible outcome.
'Free markets have allowed wealthy and powerful people to be served best,' I said. 'The government could invest in reserves of energy and water, to be used in an emergency.'

'If you tried to give everyone an equal share, the wealthy would acquire the rights of the poor, the same as in a free market,' Leblanc said. 'It would be impossible to stop them without imposing rationing.'

'A compromise is wanted, between an open slather and rigid inhumane conformance,' I said. 'How could we get people to compromise their demands?'

There was no answer.

'How can we get conservation?' I asked.

'No-one conserves anything nowadays,' said Buck, negative as usual. 'Are we going to demand the politicians change the system? Could we have a protest march or take direct action? If we waste energy and overload the power supply by having everyone turn on their appliances at the same time, they may see a need for sharing with others.'

'Hopefully we can persuade the authorities without wasting resources,' I said. 'They might listen to a reasoned argument.'

'Nah,' said Buck, as if reason would not be considered. 'People won't conserve until they are about to run out.'

'Everyone could be limited to a certain quota they must not exceed,' said Emily. She spoke with confidence, as if this was a cause she believed in.

'People taking more could pay more. That is normal for water supplies.'

'What are some other ways to encourage conservation of electricity and water?' I asked.

'Next reader, please.'

'High charges for energy help conserve it,' said Sophie laconically.

'She's right,' said Kelly. 'People can buy as much as they want at present. High charges discourage them from being greedy. Could electricity and water be taxed?'

'Thank you for that suggestion, Kelly,' I said. 'Most of the energy we use is taxed. When we pay the electricity account and for petrol, *there is a hefty tax payment. Are you saying the current charges are not high enough?'*

'Maybe.'

'A lot of people would agree,' I said. 'Our next lesson will be about consuming electricity and water at home. Would you prepare by reading the next section.'

At the next lesson, we began discussing my list of questions.

'Do people have a right to waste energy as they see fit?'

'We don't ration water or petrol, although people can go on trivial journeys with only one occupant, the driver,' said Kelly. 'It's the same with water. People waste it.'

'Could Government leaders declare rules to reduce waste.'

'Government workers should set an example,' said Tracey.

'A few people make a lot of money out of waste. If it were otherwise, it would not be waste. Why are empty office buildings lit up at night?' said Leblanc. 'Presumably the tenants are paying for the electricity, but to what end, outside office hours? It could be a rort!'

'Energy and water are rackets,' said Tracey. 'A lot of energy and water are being wasted. Depending on energy and water prices, some people are ripped off, while others get a bonanza.'

'We need a central planning authority to expose the over-consumption of some consumers.'

'The number of bureaucrats checking up on bureaucrats should be a minimum,' said Leblanc. 'The level of bureaucracy next above or below the position you are considering is always blamed for rackets. The solution is to have less bureaucracy.'

'How could governments be prevented from pursuing self-interest?'

'I could encourage you to uncover what politicians are doing,' I said. 'I am a passionate conservationist, but it is not my job to rouse juvenile eco warriors. You are supposed to be learning how democracy *can work peaceably. We can discuss later whether any direct action is necessary.'*

'The rorting of resources has gone on for far too long,' said Buck.
'Yeah,' chorused Sophie, Biggs and Norman, our other radicals.
They were in a rebellious mood and I hoped I could keep a lid on it.

TASK 5
What uses of energy and water would be the last to be reduced when supplies are scarce?

TASK 5 Answer
Energy: electricity for hospital emergency departments.
Water: for drinking.

CHAPTER 6 CONSERVING WATER

How do we ensure the water supply is used fairly, preventing greed and scarcity?

'In Australia, the main method for conserving water is to have to pay for it, no matter how thirsty you are, or how little money you have,' said Biggs indignantly.
'It's not fair,' said Buck. *'We have a right to our own water. If It rains on your property, the water is yours. You shouldn't have to pay for it.'*
'You pay for collecting, storing and pumping,' I said. *'You get your rights by sharing in costs.'*
'Water should be more expensive,' said Kelly. *'In the drought in Brisbane in 2007, household consumption halved. People should be discouraged from wasting water. When the drought ended, people's consumption was restored to the previous level, showing that half of the water demand is superficial.'*
'Next reader, please.'

Organisation of supply requires authority and legislation. Outages of electricity or water are unusual. Paid electricity and water are normally available on demand. When there's a shortage, there is a problem. It has been managed by governments but private companies have bought the businesses.

'Are people protected from greed?' Biggs asked. *'Have the private companies a licence to print money?'.*
'There are regulatory agencies,' I said. *'I don't know how effective they are.'*

'Economic growth increases usage of energy and water,' said Sophie. 'Governments and corporations have a growth mantra.'

'What's a growth mantra?'

'It's like the words in a chant,' said Sophie. 'They keep saying: 'We are growing' as if that will make it come true. The mantra 'work for a living' is used quite often too. Almost everyone believes there has to be growth. Growth was necessary when the nation was being settled but it isn't necessary now,' said Sophie.'

'It's even more necessary now, as a way of coordinating,' Kelly said. 'It enables the economy to stay in a feeding frenzy.'

'Kelly, you made that up,' said Buck scornfully. 'Stick to the facts.'

'The economy swings from hyperactivity to despair,' said Kelly. 'It's a fact. Percentage employment is like a barometer measuring growth.'

'If there's full employment, there might not be enough workers for more growth,' said Jessica, astute as always.

'A steady-state economy has been advocated and is followed in some places,' I said. 'In Bhutan they monitor happiness not growth. Economic conditions are not paramount.'

'Do they try to make everyone happy?' asked Emily.

'That's impossible,' said Biggs. 'There are people like Kelly who would want more than their fair share.'

'Biggs thinks you are greedy, Kelly,' said Buck.

'Buck,' I interrupted, 'personal attacks are not allowed in this class. Apologize to Kelly at once.'

'Kelly, you are a useless piece of excrement,' said Buck.

'Either apologise now, or go to a breakout room,' I said. Buck left.

'Emily, to answer your question, I don't know if they use happiness surveys to govern Bhutan for more happiness.'

'Wouldn't their elected candidates have to be comedians or entertainers?' said Jessica.

'Happiness is more than laughter,' I said. 'In Bhutan they surveyed nine domains of happiness: living standards, health, education, environment, community, time-use, psychological well-being, governance, and culture. The growth they looked for was overall happiness.'

'Why do Australians want only economic growth?' asked Biggs.

'All the political parties want it,' said Kelly. 'It means more workers, bigger salaries for bosses and more money to spend. The growth 'engine' is primed by immigration and credited with increasing employment and well-being.'

'Next reader, please.'

Growth uses up resources and there is an impetus for conservation. Conservation is on the conscience of the growth mongers. Conservation tinkers at the margins of supply, prolonging lives of resources. A few Green Party people oppose growth because it prevents conservation.'

'It is human nature to grab more, isn't it?' I said. 'Private dams and private power stations are allowed. People who invest in them are called conservatives, funnily enough. Wealth is concentrated and well-off people get better access to more resources.

'When there's scarcity, people are more inclined to share, at least to take less.,' said Emily. 'This happened in the drought of 2000-2007, when the dams nearly emptied. People reduced their water consumption.'

'Did everyone get an equal share in energy and water?' asked Norman.

'Fat chance,' I said. 'That would be too difficult to arrange, except with a revolution or by grabbing with force. Socialists have wanted to limit private access, with government ensuring equitable access to resources and welfare benefits.'

'We need a revolution,' said Kelly. 'Socialists want a revolution with the economy under central control.'

'Socialism doesn't work,' said Buck scornfully.

'Many people in countries like China, Russia and Cuba would disagree,' I said. 'In our mixed economy, both socialist and conservative traditions are represented.'

'Sir, are you a conservative or a socialist?' Tracey asked.
'I'm neither. I like some policies on both sides.'
'I'm going to be a conservative,' said Tracey. 'They have better parties.'
There was laughter.
'Next reader, please.'

The graph shows how consumption fell in the drought and then recovered when the dams filled. In 2007, the population of 2 million in Brisbane looked fearfully for rain to arrive and the dams on the Brisbane River emptied to 20% capacity,' I said. 'Residential water took 70% of available supplies.

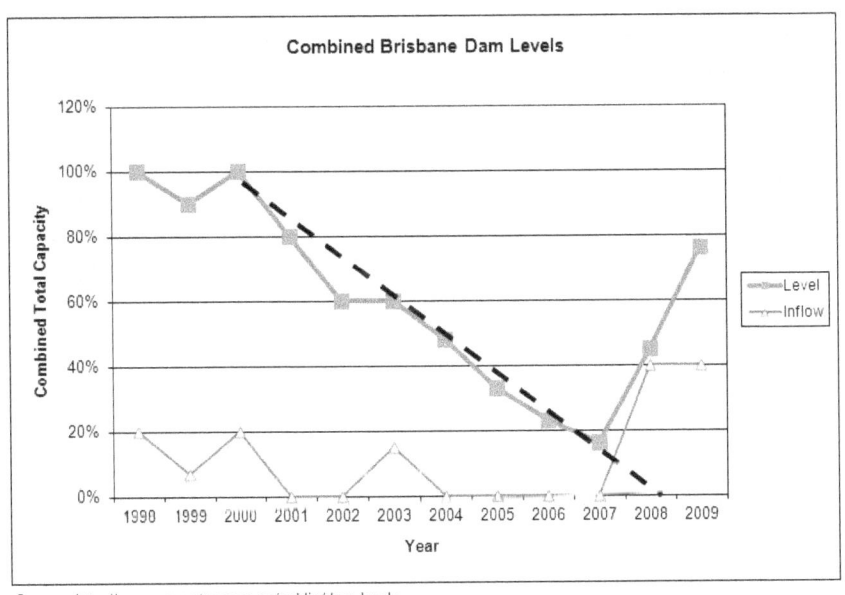

Source: https://commons.wikimedia.org/wiki/

'The graph shows how people saved and shared in the drought,' I said. 'Total water in all the dams shrank steadily. Personal consumption in 2000 was 400 litres per day, but this reduced to 136 litres per day in 2007, with a ban on watering lawns and plants, banning of hosing, no grey water recycling, low-flow shower heads, banning of washing vehicles, restricted use of pools and spas, less frequent washes, installation of rainwater tanks, turning off the water while shaving, installation of dual flush toilets.'

'There was water saving when disc jockeys played four-minute 'shower songs', so residents could track the length of their showers,' said Tracey.

'We thought the world would come to an end if we couldn't take a 15-minute shower. But we got used to it,' said Kelly.

'There was a strong feeling of community during the drought. People greedy with water were criticised in public,' Pamela said. 'People informed the government of illicit hosing and watering of gardens.

'The drought affected livelihoods too,' said Leblanc. 'By mid-2010 the Australian Government had paid approximately $4.4 million in direct drought assistance to affected farmers.'

'The graph shows there is a system for conserving and sharing scarce water,' I said. 'There may not be a system for conserving and sharing energy, but following the oil shock in 1973, shortage has been avoided by commodifying crude oil trade in a free market. The oil market regulates supply. It has enabled better forecasting and smoother trading.

'We'll end there. For next time, answer the question below.'

Kelly and Tracey lingered, chatting together. Friendly alliances added depth to classroom discussions.

TASK 6
How can water be conserved when supplies are scarce?

TASK 6 Answers
Water can have prescribed hours of usage for voluntary activities, such as showering. Certain activities can be banned e.g. hosing.

CHAPTER 7 REDUCING ENERGY AT HOME

Buck arrived and was the centre of attention.

'I apologise,' he said quietly to me.

'For?' I prompted. I had forgotten sending him to the breakout room.

'I was rude to Kelly.'

Sophie clapped. 'Good on you, Buck.'

'Apology accepted.' I said. 'Buck you can sit down.'

Biggs had his hand raised and waited until he had our attention.

'We can't use less energy, or we'll be shivering and starving,' he said.

'You can leave out watching TV and playing games,' said Norman.

'First reader, please.'

'There are many ways to reduce energy and water consumption at home. Costs for electricity and water have increased and we have to use less or even go without.'

'Our bodies are using energy for breathing, respiration and digestion, even when we're asleep. About a quarter of our food and warmth is taken to keep our bodies running. There is not much I can do to reduce my energy consumption when sleeping, without jeopardizing the relaxation I need.'

'I suppose my old stomach churns my food continually,' said Biggs. 'If I eat regularly and take time to relax after a meal, perhaps I would use less energy.'

'We waste a lot of food energy even before we eat it,' I said. 'Planting, cultivating, harvesting, transporting, storage and refrigeration all can reduce the energy in foods we consume. Foods with minor blemishes or the wrong shape are often discarded. I have heard that a third of all food that goes into refrigerators is wasted.

'Keeping food fresh is difficult. Supermarkets routinely move food to the front of the display when it nears its use-by-date, so it will be taken first. I always take from the back, where it's freshest.

'Eating raw foods such as salads can save energy. Peeling things, carrots, potatoes and the like, can lose valuable roughage.

'There are many ways to reduce energy consumption at home. it is easier to relax in a dim place, but I need bright lights to see what I am doing. How many lights do you have at home? Does anyone regularly use more than 50 lights?' Supposing you use 20 lights for 5 hours daily and each is 10Watts, that is 1000 kilowatt hours, costing 40 cents per day at the peak tariff rate. It adds up over the months.

'Does it make sense to reduce home lighting?' asked Emily. 'Is the risk of eyestrain and accident worth it?'

'I suppose every situation needs enough light but not too much or your eyesight could be strained,' I said. 'When I leave a room, I switch off the light. The saving is only a few cents per hour, but I reduce consumption by switching off when I can. The additional wear on light switches may not cost much, but cutting demand reduces installation cost of electrical generators and dams that despoil the countryside.'

'You're joking,' said Leblanc. 'The cost of using a light switch isn't enough to make any difference.'

'It's not much trouble and adds up,' I said. 'Switch by switch it doesn't amount to much, but I apply it to all switches and it makes a difference to my power bill. A light left on without anyone present in a room is likely to be switched off at my place. Some homes are low lit and their occupants are content to dine and recreate in gloom. It is better for watching television but not so good for reading.'

'Is much saved by switching off?' asked Jessica.
'Next reader, please.'

'The cost of our use of electricity at home can be informed by the cost consequences of using appliances. Consider the cost of warming a room. According to ChatGPT, heat to warm air in a nominal room 6m X 6m and 3m high, up to 200°C is 6.66 kWh per day and would cost me $6.33 per day.

There is an additional cost, of causing global warming. The heat disperses, being conducted away through the walls, warming the neighbourhood significantly. City temperature forecasts are highest in industrial and built up areas. The phenomenon is known as a 'heat island' and these are places with most global warming. The environmental cost would probably be less than my cost of the electricity, but when we use electricity we should reckon that there is a cost to the community. In the future, electricity companies could charge their customers for environmental effects and pay compensation to climate change victims.

'The greatest electrical loads in homes are usually for heating rooms and water,' I said. 'People vary in their preferred temperatures. Body temperature of 37.6°C is comfortable in cooler air. Air conditioning and fans keep room temperature down to a comfortable level of about 20°C. Reverse cycle aircon and radiant heaters can warm the air to around 22°C.'

'Putting on and taking off garments is the best way to adjust body temperature indoors,' Jessica said.

'Unless you are friendly with the people you are with, stripping off may not be acceptable.

Next reader.

'In countries like the UK and Canada layers of clothing are used to adjust between locations inside and outdoors, in Australia the same clothing goes inside or out, with the aircon adjusted for comfort.'

'In our first chapter, we looked at a table of per capita electricity consumption in the nations. The high users were countries with cold winters. But Australia was one of the highest users,' said Tracey. 'Why is that?'
'Next reader, please.'

'Although Australia is warm, perhaps Australians are accustomed to controlling their home temperature environment, they are relatively intolerant of climate warming,' I said. 'Removing clothing to allow evaporative cooling of perspiration, or adding garments for warmth, could help Australians cope. Uncovering skin for cooling could allow irradiation. Australians are wary of uncovering in the sun.'

'Not every Australian home is heated or cooled the same,' said Kelly. 'A home can be as cosy as a kitchen or as cool as a monastery, when different attention is given to the comfort of the occupants.'

'The rest of this chapter will be about energy and water used for cleaning, washing and domestic appliances,' I said. 'Would you make a list of changes to these routines in your home, that would reduce electricity and water use. What is the single change that would bring most benefit?'
Emily Barton had her arm up.
'Yes Emily?'
'The change of most benefit in our house would be the saving from using the washing machine, dishwasher and oven during peak hours, to the off-peak tariff between 10 pm and 7 am.'

'That's right,' I said. 'Well done. The appliances would use the same amount of electricity, but the tariff rate would vary with the time of day, up to 50% lower.'

'Would they use less electricity off peak?' she asked.
'No. The off-peak electricity is just priced cheaper.'
'But it would be an important saving – about half the cost.'

'Yes, absolutely. By timing your electricity use at off-peak times you will help conserve the system's generating capacity and save money too. Everyone is better off.'

'Getting out of bed to run appliances at night would be a drag,' said Kelly.

'Washing machines can also be managed to reduce energy and water use,' I said. *'You can lengthen the life of the machines using short cycles and low temperatures. Now you'll do an investigation to test different machine washing cycles. You'll do a small scale test to simulate a washing machine. You'll compare dirt removal from pieces of white cloth, soiled with either chocolate, ketchup and grease, shaken in closed jars with half a teaspoon of various washing powders, including cold power. Wash identical samples with cold and hot water, shaking each jar by hand 10 or 30 times. Repeat the tests without using detergent.'*

The students had brought equipment for the test and were able to work separately at home, showing each other their shaking techniques, enjoying the agitation activity.

There was a complication when the screw cap came off a jar while it was being shaken and the contents went everywhere.

The experiment found that washing in cold water and cold power detergent, with 10 shakes, was best, except for grease which was hard to remove and came cleaner with hot water and cold power detergent.

I explained to them that the test was subjective.

'Not every detail of the method can be reproduced,' I said.
'I don't see how our test could be fairer,' said Barry.
'Were you sure which of the samples washed whiter?' I asked.

'It was a bit difficult to see any difference, sometimes,' he said. 'We needed a reflectometer.'
'A photography light meter could be helpful.'
'Why did we repeat each test with only one variable changed?
'Those were control tests, to show the effect of that variable.'

TASK 7
Summarize the findings from your washing test. Was it a fair test?

TASK 7 Answer
The test simulated a washing machine load, but the rubbing and rolling of the articles didn't scale down accurately. The tests were probably less energetic and the findings couldn't be assumed to apply to full size washing.

CHAPTER 8 USING ENERGY WISELY

When washing machines were invented, they were expected to save work in the laundry, but they didn't. Laundry workers did as much work as before. Why?

'Perhaps workers felt that putting in more effort would get an even better result,' said Kelly.
'They might imagine a whiter appearance, after doing all that work.
'Next reader, please.'

The scientific method is to change one variable at a time to find out its effect. Energy consumption with cold washes was less. Electricity consumption was reduced by selecting quick cycle washes. This would extend the life of the machine and the clothing,
A long cycle was undesirable in an appliance, but usually the default set by the manufacturer was over two hours duration. That was ridiculous. Half an hour was usually sufficient. Attention to the machine's functioning could reduce the wear and tear on the machine. Domestic machines are constructed to be replaced when their material breaks down, bearings fail and they begin to leak. Long and very hot cycles could be expected to wear out a machine quickly. Care taken to reduce their loads lengthens machine life significantly. A machine performing half-hour cycles might last four times as long as one with two-hour cycles, before breaking down.

'I believe manufacturers want the appliance to wear out so they can sell another,' I told students.
'Would they really wear them out quickly?' asked Norman.

'I don't know,' I said, 'but they wouldn't risk their brand getting a bad reputation from cycles that were too short. Appliances have their obsolescence planned. Replacement can be necessary when something as trivial as a hose breaks. Cost of a serviceman to replace a hose can cost as much as a new machine.'

'Is that likely?' asked Kelly. He was the class's cynic.

'For an old washing machine soon to be replaced, the call-out fee can be $300, a new hose $150. Getting a new machine costing $650 could be better value.'

'I would replace the hose myself,' said Kelly.

'Good luck with that,' said Jessica. 'The bastards have made it necessary to take their machines apart even for small repairs, requiring special tools. '

'Next reader, please.

'The government tried to get me to conserve energy by installing a Climate-Smart Blue Tooth meter beside my power box, to advise me of the cost of operating my appliances.

How would you expect the cost of using my appliances to compare, from most expensive to cheapest? Here they are, not in descending order of cost.

WASHING MACHINE
DISHWASHER
TUMBLE DRYER
OVEN
MICROWAVE
RADIANT BAR HEATER
HOT WATER SYSTEM
AIRCON COOLING
AIRCON HEATING

What would be the running costs, in descending order? The government's intention was to draw attention to the high cost of using appliances, so our use of electricity would be frugal.'

'The aircons and oven would cost most,' said Biggs

'The oven was most expensive, but the aircon was among the least expensive,' I said. 'I can't remember the order of all my appliances.'

'The microwave would be cheapest,' Norman said. 'They don't heat up much.'

'You could be right, Norman,' I said. 'It is difficult to compare like with like. 'My aircon was cheapest. Hot water was expensive, so I began to use cold water for washing and for brief showers.'

'The idea of the smart meter was for consumers to learn the costs of using their appliances,' said Norman.

'Some appliances and devices had their energy usages labelled, allowing comparison between brands.'

'Next reader, please.'

'Not many people look at appliances' labels before deciding which to buy,' I said. 'They don't make rational decisions about conserving energy. Most of the time they simply use their appliances and pay the bills. The electricity supply companies don't make appliance usages transparent or invite customers to rationalise appliance settings.'

'We had a Climate –Smart meter in our house too,' Tracey said. 'We were surprised at how cheap the aircon was, so we began using it more. The Government's campaign to conserve electricity back-fired at our place and our electricity bill Increased.'

Everyone laughed.

'Small daily reductions in appliance use can add up to a significant saving in a monthly electricity account of $400,' I said. 'I use my electric oven when the smart meter tariff rate is 'on the shoulder', costing 73% of the peak rate. Off peak is 60%.

'It doesn't sound like much fun waiting around for cheap electricity,' said Kelly.

'Yeah. Get a life,' said Buck

'Life isn't all fun and games,' I said. 'People save money when they can. I'd like everyone to make a list of ways they can save money with energy and water, for next lesson.'

TASK 8
List ways to save money with energy and water.

TASK 8 Answers
Install a time-of-day meter.
Use appliances at off-peak and shoulder demand times.
Put on garments or take them off to control body temperature.
Use fully loaded short washing machine cycle.

CHAPTER 9 APPETITE

'Thank you for your lists of energy and water saving ideas, everyone. Did anyone show them to a parent? You could suggest they share with you any saving they get.'

'Next reader, please.'

'We have seen earlier that our bodies convert food so we can use it. In this chapter you will find out how your appetite chooses how much you eat. Besides energy from eatables, you may have an appetite for goods, appliances and conveniences for your comfort. You could borrow other people's. One day you could acquire an air conditioner, a television and a washing machine of your own. You could want transport, to travel around your territory, requiring you to own a fast and luxurious car. In the years ahead, your appetite for material things could take most of your earnings.

'Is it correct to say you have an appetite for certain things?'

'Appetite is regulated by various hormones. Satiety is reached as energy homeostasis, a state of balance among all the body systems needed for the body to survive and function correctly,' said Leblanc, reading from his phone.

'What are hormones for?' asked Kelly.

'Hormones are chemical messengers that tell your body what to do,' I said. 'They control organs such as the stomach, in response to hunger. They regulate metabolic needs which change body size and organ size.'

'Do the hormones control body size and shape, by stimulating desire for appealing foods?' Jessica asked.

'There doesn't seem to be anywhere in your body that keeps a 'body size map' which individuals try to override as they struggle to lose weight,' I said. 'There used to be an idea that if you were born with a self-concept of fat, you have to suck it up. Nowadays it is recognised that body shape can vary and people can change their shape, to some extent.'

'Appetite for many people controls their weight and body shape,' I said. 'Appetite is a wild stallion to ride. Now, what do you think I mean by that?'

'It is difficult to achieve the body you want by controlling your eating,' said Biggs.

'Many people struggle with appetite,' I said. 'Controlling your eating is difficult unless you are used to it. There are conflicting reports of how painful starvation is. Hunger strikers describe chronic pain but a medical document details how a very fat man in the USA had water alone for over a year. He lost half of his body mass. If he could lose so much weight, you could lose some of yours, if you need to.'

'It's hard to believe he could starve for so long,' said Kelly.

'His body consumed his fat. In another case of starvation, James Scott, a medical student from Brisbane, became lost in the Himalayas and had no food for 44 days, until he was rescued,' I said. 'When I asked him about his ordeal, he did not mention any pain. He spent nearly all his time curled up in his sleeping bag in a cave. He got up once a day to pee and drink melted snow. When they found him, he had lost 25 kilograms and was near death, with his fat almost all consumed. Fortunately he had been in good condition with some body fat when he became lost. Energy was needed to stay alive and his body was eating itself.'

'Are anorexia and bulimia voluntary starvation, caused by bad self-concepts?' asked Sophie.

'Possibly. They are illnesses, caused by dysfunctions.'

'Are they caused by dysfunctional appetites?' she asked.
'Possibly. Some people eat as a social activity,' said Tracey.
'I agree. There are many possible reasons for appetite.'
'Can we do a food tasting test?' asked Kelly.
'Next reader, please.'

'The Marshmallow experiment was a famous psychology study conducted at Stanford University in 1972. It tested delayed gratification, They wanted to find out if some individuals are more able to control their appetites than others. They gave children a marshmallow and told them that if they hadn't eaten it when they came back, they would get another one. What do you think the children did?'

'Some of them ate it immediately?'
'Yes. Other children chose to delay gratification to receive greater rewards in the future.
'Next reader please.'

'They followed those children's performances through school and university and found that those able to delay their gratification were most successful in their work, at play and socially. Conversely those unable to delay and ate theirs, were unsuccessful. The researchers followed each child for more than 40 years and over and over again, the group who waited patiently for another marshmallow, or some other food, succeeded in whatever capacity they were measuring. In other words, the experiment found that the ability to delay gratification was critical for success in life.'

'I get it,' said Kelly. 'We don't need to do that experiment.'
'Kelly, I'll bet you couldn't delay gratification,' said Tracey.
'Hell no,' said Kelly. 'I live in the moment.'
'Does the experiment show patience is a virtue?' said Tracey.
'I think so,' I said.

'Another type of gratification,' I said, 'is price elasticity when a consumer has to buy something at a high price, or wait for the price to decrease. If their demand is inelastic, their demand remains relatively constant when prices change up or down.'

'Is the price for electricity elastic?' asked Jessica.
'You don't usually get an opportunity to negotiate electricity price. It's take it, or leave it; that is, inelastic.'
'Next reader, please.'

'Demand for both food and energy may be relatively inelastic when the need has traditional or status significance. Conversely, demand for food and energy commodities can be negotiated in free markets.

'There is international commitment to valuation of commodities in markets. The assumption is that willing buyers meet with willing sellers, with traders who represent the wants of their people in negotiations conducted transparently and fairly. When buyers or sellers change their conditions, there can be large changes that affect people in other countries.

'Food and energy commodity prices control household budgets. Market price changes can affect purchasing and soon disrupt consumption and diets. For example, when the prices of meat and coffee increase, people buy substitutes. But the price of gasoline does not affected demand as much, because drivers can't buy substitute fuels and regard most journeys as inelastic. They can't delay their journeys.

'Diets flex to economise when prices of foods change,' I said. 'In a lesson soon, you will plan a diet that balances your energy inputs with your activity outputs. When prices change, brands may be substituted without invalidating the energy balance.

'Can you see how our appetites negotiate our diets. It's the same with other requisites that we 'marshmallow'.'

TASK 9
Give examples of an appetite that cannot be delayed when there is a supply failure.

TASK 9 Answer
Electricity for critical hospital surgery. Recharging emergency lighting. Drinking water.

CHAPTER 10 TRAINING FOR PERFORMANCE

'Whether you are a performer in athletics, sports, on the dance floor, at schoolwork, or another skill, to succeed you will need to train and practice, with utmost exertion to achieve your goals masterfully. There are several basic training methods for gaining physical and mental strengths. We will look at some ingredients and how you could use them. This is not a recipe like how to bake a cake, because everyone is different.

In this chapter we will start your mental training. To use your energy well, your mind needs to be in control. You need also to control what you eat. We're going to prepare for dieting next lesson.

'Is this psychology?' said Sophie.

'Yes.'

'It's not science,' she said. 'It's not real. It's in our heads.'

'We can't observe what's in our heads but we can make reasonable inferences,' I said. 'You will find the concepts are real enough.'

'Freud inferred that thinking was mainly about sex,' said Sophie.

'You can imagine it, if you want to,' I said. 'But it is a distraction.'

'When we perform are we going to be in control?' asked Emily.

'Yes.' I said. 'You are like a car driver. To drive safely and well, you need to practice driving often.'

'I have plenty of drive already,' said Kelly.

'Says who?' asked Tracey looking around.

'If you leave the car in the garage, you won't gain confidence,' I said. 'By practice and striving to drive well often, you will gain the *concentration you need to endure. Tell me, what's the first thing you do when you drive?'*

'You get in,' said Sophie.

'Yes, then you start it,' I said 'if you can. If it hasn't been driven for a while, will you be able to start it?'

'No,' she said. 'The battery will be flat.'

'How can I charge my battery then?' asked Sophie.

'You need to drive regularly to be able to start. The same with everything else, you need to keep in practice. You need to eat as you would in competition, with a balanced diet balanced to deliver the energy you will need.'

'How can I balance my diet?' asked Sophie.

'We'll find out about balanced diets in the next lesson, but here are some ideas.

'Next reader, please.'

'They say carb loading with pasta is good for running endurance. For sprinting, there are many diet possibilities, for example a low-fat, high-carbohydrate and low-fibre meal about three hours before the race, to prevent any indigestion, fatigue or stomach discomfort whilst running. You will have your own favourite foods, but remember, these are to fuel high performance. Snack foods and fast foods are not wanted.'

'What performance do you have in mind?' asked Leblanc.

'Your training should simulate your performance,' I said. 'It can be anything requiring exertion and concentration. It has to be real. According to philosopher Jean Baubrillard, reality is that which can be simulated. You will rehearse. With regular practice and a balanced diet, you will be ready to perform.

'Read on Norman.'

'Besides simulating performance and nutrition, you can overload. The overload principle is a crucial, a foundational idea in fitness. If you don't overload your body, you will never see gains in muscle strength, endurance, aerobic fitness or growth. You need to have a goal and progress towards it.

'Weightlifters use their will to cross pain boundaries and you should too. Exercise breaks down the muscle, with micro-tears that rebuild it stronger than before. A runner rebuilds her muscles every 15 to 30 days, growing smooth fibres for endurance and twitch muscle fibres for quickness. 'No pain, no gain' is the mantra.'

'Thank you, Norman,' I said. 'In the next lesson you will select your activities for a typical training day and calculate a healthy diet to balance them. Choose a goal and then plan your diet and exercise. I want you to keep to your plan for the rest of the semester. Answer the following question.'

TASK 10
Explain to someone how being ready for a goal, like a game or an exam, is like starting a car that has been left in the garage, like training your willpower to lift weights.

TASK 10 Answer
My goal is to run a marathon in 5.00 minutes. I will keep to my training activity schedule diet. I will run several marathons and also do runs at various distances, concentrating on enduring and going fast at the edge of overload.

CHAPTER 11 ENERGY IN DIETS

'In this lesson, you will find out how to plan a diet to balance physical exercise with food energy,' I said. 'You need to calculate a diet for your exercise schedule and stick to it.

'Below is an example of a person's analysis of energy inputs (Part A) and outputs (Part B) for one day, for you to analyse and conclude whether energy is balanced.

'You will balance your energy input with your output by a logical method based on scientific data. Energy is a measureable physical quantity. Do not confuse it with spiritual energy, which is unscientific and vague. If your food intake is less than the energy used in your activities, you will lose weight. It's that simple. Recording your food and exercise should be a daily habit until you are familiar with it.

'In the table below, a person calculates her percentages to be evaluated against published minima and maxima. Where more than one serving or a part serving was eaten, the total food value is shown.

FOOD INPUTS

FOOD	No servings	Protein grams	Fat grams	Carbo hydrate grams	ENERGY kJ
Boiled egg	1	5.6	5.2	0.3	302
Weet-Bix	1	2.1	0.2	12.2	252
Raisin bread	1	3.3	1.5	27.0	558
Baked beans	1/4	3.6	0.3	13.1	277
Brown bread	2	8	1.8	49.2	1016
Honey	1	0.1		21.6	365

Milk	2	15.2	17.4	21.0	1292
Grilled steak	1	31.8	33	0	915
Mushroom	6	1.4	6.3	2.4	277
B potato	2	3.6	0.2	34.2	604
Watermelon	1	0.5	0.2	7.1	126
Potato crisps	2	3.0	23.8	30.0	714
Salad	1/2	3.2	0.3	13.4	268
Fruit cake	1	2.0	6.0	18.1	537
Orange juice	1/2	0.8	0.2	14.6	239
Apple pie	2	3.3	12.4	52.6	1386
TOTAL INPUT	513.1 Food	87.5 Protein	108.8 Fat	316.8 CH	9128 Energy
	100.0	17.0%	21.2%	61.7%	
		10	20	55	
Maximum		15	35	65	

'This table has more protein than the maximum allowed and fewer carbohydrates? Does it have to be exact?' asked Jessica.

'Theoretically, yes but there can be small deviations. It can be difficult to meet all the requirements.'

TASK 11A
Answer these questions:
1. What is the total energy input?
2. What is the percentage total protein?
3. What is the percentage total fat?
4. What is the percentage total carbohydrate?

TASK 11A Answers
1. Total energy is 9128 kJ

2. Total protein is 17.0%, a little above the maximum of 15.0%
3. Total fat is 21.2% a little above the minimum of 20.0%
4. Total CH is 61.7%, between the minimum and maximum.

TASK 11B

Now calculate her total output energy from the schedule of activities below for 24 hours. The energy values are constant. Note that the total energy per kilogram is multiplied by her weight of 69 kg at the foot of the table.

ENERGY OUTPUTS

ACTIVITY	Energy (kJ) per kilogram of body mass per hour of activity (A)	Time per day hours (B)	Energy per kilogram per day (A) X (B)
Asleep	3.7	8	29.6
Basketball, moderate	26.1		
Cycling - rapidly	34.5		
Dancing, moderately active	20.2		
Dressing and undressing	8.3	1	8.3
Eating a meal	6.4	1	6.4
Football/netball	36.5		
Ironing	9.2		
Jogging	31.5	0.50	15.75
Lying still, awake	4.6	3	13.8
Playing musical instrument	10.1	0.5	5.0
Reading aloud	6.5	1	6.5
Riding in a car or bus	9.2		

Running, fast	39.8		
Standing	7.4	1	7.4
Swimming, moderately	32.2		
Typing	9.2	3	27.6
Washing the dishes			
Watching TV	6.4	3	19.2
Writing	6.4	2.0	12.8
TOTAL	NA	24	152.35
		X 69 kg	10512

Summarise the calculations

INPUT (table above)	kJ	9128
OUTPUT	kJ	10512
INPUT/OUTPUT	%	86.8

TASK 11A Contd
Answer these questions.
1. Does the food input in the table above balance the energy output?
2. What would you predict would happen if this diet and activity schedule were maintained?
3. How could you modify the diet for a better result?
4. How could you modify the energy schedule for a better result?

TASK 11A Answers
1. Input is a little below output, 86.8%
2. She would lose weight because her output 10512 kJ is larger than her input 9128 kJ.
3. She could eat more energy foods, such as spaghetti with cheese sauce. Vitamin and mineral quantities need to be considered.
4. She could get more balance by eating more or exercising less. The energy schedule could be varied daily up to the performance.

TASK 11B Contd.

Download an empty food table from above. You will prepare a table like the one above for all your ideal food inputs on a typical day from the pages of food data below, selecting your foods and copying them into it. There is data for most foods and you can estimate others. Sum the column on the right. Then at the bottom of the table calculate the column totals and percentages as shown above. Your total food energy should be under 10,000 kJ, unless your activity in the next table is greater.

The next few pages have food data to prepare your own table.

Food name	No servings	Protein grams		Carbo hydrate grams	ENERGY kJ
MEATS					
Hambrger patty	1 patty	11.4	14.8	11.9	936
Grilled steak	med avg fat	31.8	33	0	1831
Grilled steak	Med low fat	39.6	12.3	0	1176
Beef casserole	¾ cup w veg	17.0	19.3	10	1213
Beef casserole	¾ cup no veg	17.3	20.2	9.6	1272
Savoury mince	¾ cup w veg	16.5	21	10.0	1260
Fried chicken	1 avg piece	37.1	17.0	3.7	1377
Lamb casserole	¾ cup w veg	32.0	10.3	0	915
Lamb casserole	¾ cup no veg	15.3	14.4	13.6	1045
Large lamb chops	1 chump chp	10.3	17.0	0	609
Small lamb chops	1loin chop	12.0	20.7	0	995
Roast leg lamb	1 slice 80x80	7.4	11.6	0	562
Fried bacon	1 strip 33 cm	4.1	13.4	0.4	596
Grilled bacon	1 strip 33 cm	4.1	8.9	0.4	415
Pork chop	1 avg	17.4	30.4	0	1461
Sw and sour pork	½ cup	9.3	36.3	9.4	1667
Leg ham	3 slices	10.9	7.3	TR	483
Brains fried	1 set crumbd	16.2	29.5	15.5	1709
Grilled kidney	1 sheep's	9.3	3.9	TR	315
Liver fried	1 slice130x50	18.5	6.2	8.2	663
Tripe with sauc	¾ cup	16.6	6.4	10.3	714
Lambs fry bacn	¾ cup	31.5	49.2	20.4	2772
Lunch sausage	1 10 cm slice	3.7	4.6	0.1	252

Ffurter boiled	1 avg	6.5	11.9	1.0	588
Salami	1 8 cm slice	6.7	10.2	TR	508
Beef saus grilld	1 thick	8.4	12.2	5.3	709
Beef saus grilld	1 thin	5.7	8.4	3.6	487
Pork saus grilld	1 thick	7.9	12.7	7.7	701
Pork saus grilld	1 thin	5.5	8,7	5.3	508
Fish baked	1 fillet	13.4	2.3	0	327
Fish crumbed	1 fillet	18.4	14.4	11.6	1062
Fish battered	1 fillet	19.2	13.1	14.3	1045
Fish steamed	1 fillet	13.4	2.3	0	327
Oysters fresh	1 dozen	10.4	1.8	4.9	344
Salmon canned	½ cup	25.8	9.3	0	823
FOOD NAME	No servings	Protein grams	Fat grams	CH grams	ENERGY kJ
EGGS					
Boiled egg	1 medium	5.6	5.2	0.3	302
Fried egg	1 medium	5.6	9.3	0.3	453
Plain omelette	2 eggs	11.4	22.6	0.8	1041
Poached egg	1 medium	5.6	5.2	0.3	302
Scrambled egg	Egg,mlk,butt	6.6	12.4	1.8	621
DAIRY PRODUCTS					
Cottage cheese	1 tblspn	2.7	0.8	0.6	88.2
Cream cheese	1 tablspn	1.7	6.1	0.6	277
Cheddar chse	Slice50x30x10	5.2	6.6	0	336
Baked custard	Half cup	6.8	6.8	12.9	596
Custard	Quarter cup	2.4	2.4	8.0	273
Flavored milk	300mL carton	10.0	10.0	29.6	1029
Skim milk	250 mL	8.0	0.2	11.3	344
Milk	250mL	7.6	8.7	10.5	646
Ice-cream	2 scoops	3.4	7.1	13.2	541
Yoghurtflavred	1 sml carton	10.1	8.2	25.6	986
Yoghurt plain	1 sml carton	10.1	8.2	13.1	693

Butter	1 tblspn	0.1	15.4	0.1	579
Fresh cream	1 tablspn	0.4	7.6	0.6	306
SOUPS					
Crm of chickn	1 cup	3.7	5.3	7.4	386
Pea and ham	1 cup	8.7	3.7	27.4	743
Crm of tomato	1 cup canned	1.3	2.3	15.3	365
Vegetablesoup	1 cup (packet)	3.5	1.8	11.5	319
Chickennoodle	1 cup (packet)	1.7	0.6	7.3	130
Tomato soup	1 cup (packet)	0.9	TR	9.5	172
Crm chicken	1 cup (packet)	1.5	1.2	7.1	193
Pea and ham	1 cup (packet)	2.3	1.4	6.4	201
SAUCES					
Gravy	2 tablespns	0.4	5.1	6.3	306
Tomato sauce	1 tablespn	0.4	0.1	5.1	88.2
Frenchdressing	1 tablespn	0.1	7.4	3.7	340
Mayonnaise	1 tablspn	0.2	15.0	0.5	567
FRESHFRUIT					
Apple raw	1 small	0.3	0.3	13.8	222
Apricot raw	1 medium	0.3	0.1	3.8	63
Banana	1 medium	1.1	0.3	22.5	365
Grapefruit	Half medium	0.6	0.2	25.3	184
Mandarin	1 large	0.8	0.3	11.2	193
Mango	1 small	0.7	0.2	17.2	277
Orange	1 medium	1.2	0.3	14.4	247
Passionfruit	1 medium	0.7	0.2	6.3	113
Pawpaw	¼ medium	0.6	0.1	10.5	172

Peaches	1 medium	0.7	0.1	11.7	189
Pear	1 medium	0.6	0.4	21.7	352
Pineapple	1 slice	0.4	0.2	10.8	176
Strawberries	12-14 medium	0.7	0.5	8.6	155
Fruit salad	Half cup	0.9	0.3	17.6	289
Avocado	Half medium	1.7	15.8	6.3	676
Rock melon	Half medium	0.8	0.2	6.8	126
Watermelon	1 avg slice	0.5	0.2	7.1	126
Grapes	20 per serve	0.7	0.4	16.8	277
Plums	2 medium	0.7	0.1	15.6	247
Cherries	20 per serve	0.9	0.4	15.1	63.4
DRIEDFRUIT					
Dried apple	12 pieces	0.5	0.6	26.3	424
Dried apricots	5 x ½ apricots	1.1	0.1	17.1	277
Dates	5	0.7	0.2	25.3	399
Sultana raisins	1 tablespoon	0.2	TR	9.2	147
CANNEDFRT					
Cannd apricots	4 x ½ apricots	0.7	0.1	23.4	361
Canned peach	5 piece+syrup	0.4	0.1	22.0	344
Canned pears	2x1/2 in syrup	0.2	0.1	20.8	323
Cannd pineapl	2 slices	0.4	0.1	17.7	277
Cand frt salad	½ cup	0.3	0.1	20.9	331
VEGETABLES					
Asparagus cand	3 med spears	1.0	0.2	1.8	46.2
Baked beans	¼ cup	3.6	0.3	13.1	277
Beetroot cand	2 slices	0.3	TR	2.3	42

Carrot raw	1x10 cm long	0.5	0.1	5.1	92.3
Carrot boiled	1/3 cup	0.4	0.1	3.5	63
Cauliflower bld	1/2 cup	1.3	0.1	2.5	58.8
Celery raw	1 piece 15 cm	0.3	TR	1.3	25.2
Sweet corncand	¼ cup	0.9	0.3	7.3	138
Creamed corn	¼ cup	1.2	0.4	12.0	205
Lettuce	2 small leaves	0.3	TR	0.6	16.8
Mushroomsbuton	6 small	1.4	6.3	2.4	277
Green pead boild	1/3 cup	3.0	0.2	7.6	163
Baked potato	1 medium	2.0	9.1	19.7	688
Boiled potato	1 medium	1.8	0.1	17.1	302
Fried potato chip	18 average	3.4	12.7	29.3	1012
Mashed potato	1/3 cup	1.4	0.5	9.2	184
Pumpkin bld	1 avg piece	0.6	0.1	4.2	79.8
Sweet pot baked	1 lge piece	0.4	3.2	9.7	285
Tomato	1 medium	1.1	0.3	4.5	96.6
Brussel sprouts	5 sprouts	2.8	0.3	4.6	113
Cabbage boiled	1/3 cup	0.7	0.1	2.4	50.4
Squash boiled	2 squash	0.6	0.2	4.5	84
Broccoli	1/3 cup	1.6	0.1	2.5	54.6
Mixed vegetable	½ cup	3.2	0.3	13.4	268
Coleslaw	1 cup	1.85	15.25	7.9	714
Corn boiled	1 small cob	2.2	0.7	13.1	243
Zucchini	1 small zucch	0.6	0.2	4.5	84
Cucumber	3 slices	0.2	TR	1.0	16.8

GRAINS AND STARCHES	Serving size	Protein g	Fat g	CH g	Total Energy kJ
Plain white flour	1 tablespoon	1.1	0.2	7.5	151
White s-r flour	1 tablespoon	1.1	0.2	7.5	151
Macaroni boiled	1 cup	4.9	0.8	34.8	718
Noodles boiled	1 cup	6.1	2.0	34.8	756
Rice boiled	1 cup	3.2	0.3	39.0	718
BREADS					
Brown bread	1 slice	4.0	0.9	24.6	508
Rye bread	1 slice	3.3	0.7	21.6	441
White bread	1 slice	3.6	0.7	23.0	470
Bread roll flat	1 roll	3.9	0.8	25.0	512
Bread roll long	1 roll	5.2	1.0	33.2	672
Wholmeal bread	1 slice	5.2	1.5	29.9	617
Raisin bread	1 slice	3.3	1.5	27.0	558
Crumpet	crumpet	2.9	0.6	20.0	415
SPREADS					
Peanut butter	1 tablespoon	5.5	9.7	3.4	499
Butter	1 tablespoon	0.1	15.4	0.1	579
Margarine	1 tablespoon	0.1	15.4	0.1	579
Marmite	½ tablespoon	0.8	0	0.7	29.4
Vegemite	½ tablespoon	0.8	0	0.3	16.8
Golden syrup	1 tablespoon	0	0	20.2	331
Honey	1 tablespoon	0.1	-	21.6	365

Jams,marmalade	1 tablespoon	TR	TR	17.9	298
Fish paste	1 teaspoon	1.0	0.6	0.2	42
BREAKFAST					
Bran/cornflakes	1 cup	2.4	0.2	23.6	436
Rolledoatporridg	1 cup cooked	4.7	2.6	23.4	541
Weetbix	1 biscuit	2.1	0.2	12.2	252
Muesli	1 cup	4.5	23.6	27.6	720
Muesli wo sugar	1 cup	4.5	23.6	27.6	720
CONVENIENCE					
Chiko roll	1 roll	68	9.9	43.8	243
Hotdog	1 roll & frank	13.0	16.6	29.6	1348
Meat pie	1 pie	13.4	32.0	44.9	2175
Sausage roll	1 roll	22.5	32.1	22.0	1986
Cottage pie	1 average pie	28.8	30.2	46.2	2385
Hamburger	Brd roll,onion	15.3	13.2	43.6	1491
Potato crisps salt	1 small packet	1.5	11.9	15.0	714
Potato crisps lite	1 small packet	1.5	11.9	15.0	714
Frid potato chips	18 chips, avg	3.4	12.7	29.3	1012
SWEET					
White sugar	1 teaspoon	0	0	4.0	67.2
Brown sugar	1 teaspoon	0	0	4.5	75.6
Iced bub	1 avergae	3.1	4.8	33.6	793
Light fruit cake	1 slice	2.0	6.0	18.1	537
Lamington	1 lamington	3.8	7.8	33.5	898

Plain scones	1 scone	2.9	3.3	16.3	445
Plain sponge cake	1 slice	2.8	1.8	16.5	394
Iced sponge cake	1 slice	2.9	5.2	23.7	638
Trifle	½ cup	4.1	6.3	28.6	777
Cheese cake	1 avg piece	10.2	21.2	27.9	1432
Fruit pie (apple)	1 avg piece	3.3	12.4	52.6	1386
Boiled sweets	1 avg piece	TR	TR	7.3	117
Butterscotch	1 avg piece	TR	0.4	4.5	88.2
Milk chocolate	5 smll squares	1.8	6.2	11.2	453
Nut & milk choc	5 smll squares	2.8	7.6	8.9	457
Dark chocolate	5 smll squares	0.8	6.0	12.5	487
Lifesavers	1 packet of 10	TR	-	16.5	252
Peanut brittle	1 piece	2.1	3.9	22.8	541
Muesli bar	1 bar	1.6	7.9	34.8	672
NUTS					
Almonds rsted	26 nuts	5.8	17.0	5.6	789
Cashews rsted	15 nuts	5.4	14.2	8.4	722
Peanuts rsted	37 nuts	7.8	14.6	5.8	730
Almonds raw	26 nuts	5.8	17	5.6	789
BISCUITS					
Cracker, low fat	1 biscuit	0.5	0.3	2.9	67.2
Cracker, med fat	1 biscuit	0.7	1.7	6.3	180
Cracker, high fat	1 biscuit	0.3	1	2.3	84
Rye crispbread	1 biscuit	1.0	0.3	7.5	142
Wheat crispbread	1 biscuit	0.5	0.6	4.2	96.6

Tim tam	1 biscuit	0.9	5.2	13.5	428
Choc biscuit	1 biscuit	0.6	1.5	7.5	184
Cream biscuit	1 biscuit	0.9	4.8	12.8	399
Plain sweet biscuit	1 biscuit	0.5	1.0	6.2	142
Digestive biscuit	1 biscuit	0.6	1.0	6.2	147
BEVERAGES					
Soft drinks	½ large can	0	0	20.4	323
Cordials	1 glass diluted	TR	-	17.0	273
Milk	1 cup	7.6	8.7	10.5	646
Milkshakes	1 regular	10.9	12.6	31.6	1205
Thick shakes	1 medium	11.6	15.0	27.4	1230
Milo	1 tablespoon	0.8	0.7	4.1	109
Ovaltine	1 tablespoon	1.0	0.4	4.9	113
Orange juice fresh	½ cup	0.9	0.3	12.7	218
OJ packaged	½ cup	0.8	0.2	14.6	239
Pineapple juice	½ cup	0.5	0.1	15.8	247
Apple juice	1 lge glass	0.3	TR	34.5	529
Beer	1 x 7oz glass	0.7	0	6.7	403
White wine	1 glass	0	0	0.1	352
Red wine	1 glass	0.3	0	0.1	363
Spirits	1 nip	0	0	0	289

Now copy the Activity table from above, clear it and complete your energy output plan for your ideal 24 hours of activities.

TASK 11B Contd.

Compare your energy input with your energy output, completing below.

INPUT ENERGY	kJ	
OUTPUT ENERGY	kJ	
INPUT/OUTPUT	%	

Finally, modify your diet and activities to balance input and output.

CHAPTER 12 FAST FOOD

'You have found out how to plan a diet to balance activity energy output with food input energy. Fast food is quick to prepare and eat. However most fast food is unsuitable in a performance diet because it can be unhealthy. In this chapter you will consider how the fast food fad can be an unhealthy surrender, with unwanted psychological consequences.

'We'll read about this around the class, in turn.
'First reader, please.'

'Fast food is an alternative to cooking and eating at home. We let advertisements for fast food intrude into our lives. This is where we go wrong.

'Recall a fast food advertisement with a young actor who makes a social gaffe at the office, or at home, with friends or family present. All eyes are on him as he squirms with embarrassment, until he says, 'Did someone say KFC?' and immediately, a picture of Kentucky Fried Chicken's garish food pops up, the tension is relieved and the group quits what they are doing to gobble greasy sugar-laden fast food. A percentage of viewers seeing this ad would be prompted to buy KFC. Fast food has become a social icon lacking reason, propelled to fill awkward moments with tradition. The supplier wants fast food consumption, having nothing to do with nutrition, nor value for money, nor societal effects. Consumers have little interest in the quality of the food and buy it for its reputation and ready availability.'

'Is that scene from a KFC ad?' asked Kelly.
'Yes. I am using KFC as an example,' I said. ' It is typical of how fast foods are advertised.'

'Read on, please.'

'KFC advertisements employ fakery that is dishonest, demeans viewers and diminishes respect for KFC as a responsible enterprise. Gaining public respect doesn't pay them as well as deception. They apply an ethos of responsible corporate citizenship that is a tissue of lies.
'Fast food buyers are often young people in a hurry. Their diets are determined by fashion, emotional appeals and peer pressure, without personal preference, or diversity. They do not discriminate between foods. At home they probably do not have opportunities to select foods, nor do they have preparation choices, nor can they choose meal quantities, nor dietary balance. They expect to get the same fast food as last time.'

'Do they always get the same as last time?'
'No. Young people like to experiment.'
'Read on please.'

'Fast foods appeal to hunger, because advertising elicits desire to eat,' she read. 'They see photos of colourful foods being enjoyed by beautiful people. When one of them recites the mantra: 'Did someone say KFC?', or alternatively, when they hear the thumping rhythm 'I don't care!' consumers' imaginations are captured by desire to obtain this food and eat it. It would take a strong leader to divert the salivating group to other brands, or to, say, Japanese food. They want instant gratification and resist surveying of members' preferences or shopping around for better food.'

'Is this saying customers grab the food and gobble it?' said Tracey.
'Customers wolf it down,' said Biggs. 'The next bit explains.'
'Read on, please.'

'Eating fast food is presented in advertising media as an orgasmic reflex orgy with gluttonous satiation readily available,' read Jessica.

'Fast food is shovelled down in an uncontrolled frenzy. KFC ads play the hunger emotion for every finger lickin' mouth waterin' sensation they can evoke, rejecting the possibility of refinement in eating and mealtime talk. Fast food is consumed quickly, in large pieces, chewed with open mouth, wiped on the back of the hand.

'The suppliers want eating to be fast, increasing sales,' I said.
'Read on, please.'

'The purpose of sport, reality TV, game shows, sitcoms, news, weather and all popular entertainments, except reading, is for corporations like KFC to take money from buyers of: fast foods, branded merchandise, music, movies, clothes, groceries, cars, holidays, devices and toys,' Jessica read. 'The profit ends up in investors' bank accounts. Corporations present their products, at venues and on media, competing for the best promotional and selling opportunities.

'Is fast food part of a presentation that entertains people, while taking their money?' asked Leblanc.
'That is the interpretation I make,' I said.
Read on please.

'Buyers face layers of images promoting the food. Fast food advertisers appeal to popularity. The easiest way to convince somebody to buy a product or service is to prove that everyone else is buying it. Once something becomes a widely recognized phenomenon or a trend, it is assumed it must have merit – otherwise it wouldn't be so popular. Right?
'This is sometimes called the bandwagon effect or Matthew Effect. 'The Bible says: *'For whosoever hath, to him shall be given, and he shall have more abundance: but whosoever hath not, from him shall be taken away even that he hath.'* Matt 13:1
'In other words, novices copy influencers. The people who scoff trendy fast foods are 'with it'. The right fast food can make you

friends. The more popular something becomes, the more people buy into it and consequently it becomes even more popular.'

'In the book industry, this is referred to as the 'bestseller effect,' I said. 'Buyers do not exercise personal taste or aesthetic sense. They buy bestsellers. KFC have actors and rent-a-crowd falsely to show social approval of their products.

'Fast food vendors create venues and contexts to stimulate consumption,' she said. 'The corporate focus on profits can modify an event, imposing audience timeout for peeing, scoffing fast foods and buying merchandise. Popular entertainments require only grunts between mouthfuls, tears and hilarity. With a full mouth, standards of athleticism, observation of rules and aesthetics, cannot be discussed either reasonably or critically Audiences merely cheer, push, applaud punch-up, disdain courtesy and dispute referees' decisions.

'Fast foods are sold fast, delivered fast and eaten fast,' I said.
'Continue reading, please.'

'Jumbo tubs of popcorn condition audiences with simultaneous eating, as a sublimation of the competitive tension of the event they are watching. Packaging enables eating with the hands, without plates, knives or forks. There is bonding between strangers eating the same fast food. In this way, audiences at public and private venues have been taken over by fast food consumption with conspicuous waste of packaging, uneaten condiments, sauces and sugar. Our fast food culture will be condemned in history by the need to dispose of mountains of leftovers, contributing to the burden of excessive resources usage and pollution.'

'The consequences of fast food are ill health, physical and mental, from over-indulgence, abandonment of self-control and mindless consumption during media-presented entertainment. Hopefully the fast food craze will disappear as suddenly as it appeared.

'The worst result of fast food is thoughtless followership and surrender of discrimination. Fast food emporia thrive on instant gratification. A leader of a group, seeing a fast food advertisement, asserts her authority: 'Shut up and take my money!'. It's an easy sell.

When her group walk into a fast food depot, they are pushed into buying ancillary products, paying for them immediately, sight unseen. The food is already prepared and it is a short wait until the package is delivered for immediate consumption. The food cycle, which began with seeing an advertisement, includes receiving the package, opening it and comparing with expectations. It culminates in eating until bloated, then sinking into a torpor, as their digestive systems grapple with the toxic shock of fat and sugar overload. They stumble away, bleary-eyed, to look for some innocuous entertainment to numb their brains, or a casual crime to commit.

'We cannot expect the quality of fast food to improve,' I said. 'Like many industries, fast food suppliers are self-interested and greedy, in the capitalist mould of cultivating indiscriminate unhealthy over-eating. It is not so widely known that the purveyors' anti-societal intent is to create an addiction to fast food. This harms many people. I hope you will quit eating fast food and get more enjoyment and better nutrition from your food purchases.'

'Thank you for reading.

TASK 12
List advantages and disadvantages of fast food.

TASK 12 Answers
Advantages: Fast food provides food energy, of a sort.
Disadvantages: Neglected are nutrition, value for money, and societal effects. Eating lacks self control. The advertising is emotional and uninformative. Fast food induces consumers into a passive acceptance mode in which they bond with others eating low quality fare.

CHAPTER 13 FOSSIL FUELS

They are called 'fossil' fuels because they formed from plant and animal remains buried under sediments long ago and preserved as fossils. Pressure and high temperatures transformed them into fuels convenient to use, as either gases, liquids or solids. Oil, natural gas and coal have fuelled electricity generation and industrial furnaces, keeping electric motors and industrial processes going. Because they are blamed for climate change, their use is supposed to be declining but their replacement in many powerhouses worldwide is not imminent despite the exhortations of international treaties and armies of protesters. When necessary electrical storage is included, renewable energy is more expensive than old fossil fuel stations shut down.

'Australia possesses about 21.6% of the World's coal reserves but most of Australia's oil and gas discovered have been depleted or exported.'

'In this chapter I will relate my experiences when working for a coal company.

'First reader, please?'

'Japan has little coal or oil and until recently has wanted Australian fossil fuels. I showed visitors from Japan the site of a coal mine proposed on the Darling Downs. They wanted to import coal to convert to oil and increase their energy security, which in WWII had been Japan's Achilles' Heel. They had lacked fuel for their planes and ships.

'When I visited Japan my guide showed me where there had once been a forest on a mountainside.' He said: 'During World War II, the Americans blockaded ships bringing oil and we could not fuel our war machines. We dug up the pine trees, cut off their roots and distilled turpentine from them to use in our aircraft engines. Doing this was hurtful because Japanese people love their forests.'

'My job with Wattle Mines was to develop export of coal to Japan and to other countries. I travelled to mines in Australia and overseas. At that time the Japanese wanted to secure their nation's oil supply and were developing technology to convert Australian coal into oil. Our team was planning to develop a new coal mine, on the Darling Downs, to supply a conversion plant at Sumitomo's steel works at Kashima, north of Tokyo.

'Jessica, read on please.'

'We planned an enormous project, with coal being railed to Gladstone for export by sea to Japan. The vast coal resource would be mined by huge draglines. Washing stone from the coal was a challenge and we planned to bring water to the mine from a dam to be built on the Dawson River near Theodore. Water was scarce and agriculturalists opposed the dam and the mine. In Australia, allocation of water is always a political process.

'Processing coal to oil takes water,' I said. 'Unless the government gets behind it, it won't happen.'

The state government was enthusiastic and promised water. Japanese visitors inspected the site of the proposed coal mine and approved it.

'You have only 10 metres of rock above 20 metres of coal,' said a visitor to the open cut, viewing the outcropping seams of coal.

'That's correct,' I said.

'It's amazing,' he said. 'In Japan, coal mines have hundreds of metres of rock over less than a metre of coal.'

'Although Australian coal was cheap, oil could be imported from the Middle East more cheaply than it could be converted from our coal,' I had said.

'We could build a coal-to-oil plant to extract 100,000 barrels per day of oil from Australian coal,' the Japanese visitor told me. 'It would be a massive plant, costing billions of dollars, with a workforce of 5000 for many years. On the other hand, the Saudis can drill, before lunch, one oil well that would produce an equal quantity of oil, 100,000 barrels per day.'

'Relations between the Saudis and Tokyo improved and It made our project seem futile. Japan's interest disappeared.'

'There was also lowered interest in coal because pollution was regarded as greater than from oil,' said Jessica.

Tracey interrupted. *'Renewable energy now has most interest for electricity supply but no substitute has become available for aircraft fuel.'*

'They could harness skeins of swans and geese,' said Buck, guffawing. There was laughter.

Read on, please.

'There is still demand for Australian fossil fuels to export to China, India and Indonesia. These countries may not be able to afford to generate enough electricity without using fossil fuels, despite worldwide limitation on emissions.'

'Natural gas could generate electricity with a carbon footprint only half of coal's,' I said, 'but methane leakages are worse than CO_2 and overall its footprint could be as bad as coal. It can be piped and stored, but the problem is that there isn't much gas remaining in the ground and most of it has already been sold for peak power generation or sold overseas. The gas remaining is often in a 'tight' geologic formation, requiring fracking to coax it out of the rock.

'The future of fossil fuels in Australia seems bleak, despite the huge reserves, because of government restriction of fossil fuel use.'

TASK 13
When German imports of oil were blockaded in WWII, how did the German armed forces fuel their aircraft and tanks?

TASK 13 Answer
Coal was converted to oil, by hydrogenation, a costly process.

CHAPTER 14 ENERGY CONVERSION

Various devices have been used to convert energy input to energy output. Complete this table.

ENERGY IN	ENERGY CONVERTER	ENERGY OUT
electrical	storage battery	chemical
electrical		sound
	hot air balloon	
chemical		kinetic
chemical	steam engine	
kinetic		sound
chemical		electrical
	electric radiator	
kinetic		electrical
	solar panel	
electrical		kinetic

The output of each converter is less than the energy it takes in.
The first law of thermodynamics is that energy cannot be created or destroyed, only stored or converted. Efficiency % is output/input x 100. If efficiency was 100%, there would be perpetual motion, which has never been achieved anywhere.

TASK 14A
Complete the table above. Answers are below.

TASK 14B

Energy conversion efficiencies for converters and their energy chains are listed in the table below. Systems having more than one converter are called energy chains.

Not all energy conversion is efficient and since the Industrial Revolution, many converters have been used which are very wasteful. Some of the inefficient technologies have been superceded, but others are surviving. The table below shows that the efficiency of conversion of energy can vary from 99% to 2% in various devices.

Use the information in the table following to answer the questions below.

CONVERTER	ENERGY CHAIN	EFFICIENCY %
Electric generator	Kinetic→electrical	99
Electric heater	Electrical→ heat	99
Large electric motor	Electrical→kinetic	92
Dry cell battery	Chemical→electrical	90
Home gas furnace	Chemical →heat	83
Storage battery	Electrical→chemical→electrical	73
Home oil furnace	Chemical→ heat	64
Small electric motor	Electrical→kinetic	62
Liquid fuel rocket	Chemical→heat	48
Steam turbine	Thermal→kinetic	45

Steam power plant	Chem→heat→kinetic→electrical	41
Diesel engine	Chemical→heat→kinetic	37
Aircraft gas turbine	Chemical→heat→kinetic	35
Auto I.C. engine	Chemical→heat→kinetic	25
Fluorescent lamp	Electrical→radiation	21
Solar panel	Radiation→electrical	20
Steam locomotive	Chemical→heat→kinetic	8
Incandescent lamp	Electrical→light	2

TASK 14B Questions
1. Work equivalent to 15 megajoules is to be done by small electric motors. How much electricity must be input?
2. What is the efficiency given for a large electric motor compared with an automobile combustion engine?
3. What is the energy chain of devices for a steam power plant?
4. Fuel containing 100 megajoules of energy is placed in a diesel vehicle. How much energy would be produced by the engine?

TASK 14A Answers

ENERGY IN	ENERGY CONVERTER	ENERGY OUT
electrical	storage battery	chemical
electrical	**loudspeaker**	sound
heat	hot air balloon	**gravitational**
chemical	**rocket**	kinetic
chemical	Steam engine	**kinetic**
kinetic	**drum, guitar**	sound
chemical	**battery**	electrical
electrical	electric radiator	**heat**

kinetic	Generator, dynamo	electrical
radiation	solar panel	**electrical**
electrical	**Electric motor**	kinetic

TASK 14B answers
1. 15/0.62 = 24.2 MJ
2. 92% and 25% (maximum)
3. Chemical→heat→kinetic→electrical
4. 100 X 0.37 = 37 MJ.

CHAPTER 15 BIOENERGY

In this chapter, you will explore energy from green plants. The problem with fossil fuels is that they produce carbon dioxide when combusted. Carbon dioxide can grow plants to produce plant sugars for bioenergy.

'Could bioenergy help to fix the fossil fuels problem?' asked Emily.
First reader, please.

It may seem that the carbon dioxide from combusting a fossil fuel could be removed by growing sugar cane to produce vehicle fuel. But it would not replace fossil fuel entirely. The 'must' stage in fermenting sugar can only reach a concentration of about 15% alcohol before the yeast dies. It liberates only part of the carbon dioxide in the plant sugars. To raise the alcohol content to a concentration where it will allow combustion in engines, anaerobic fermentation is required, which produces more carbon dioxide. The process merely recycles the carbon dioxide used in cane growth by photosynthesis.

'Does that mean bioenergy isn't a remedy for fossil fuels?' said Jessica.
'Correct. We will see why presently.'
'Next reader, please.'

'Bioenergy can be obtained from plants which grow back or are replanted, without consuming the supply, as occurs with fossil fuels. During the 2022 crop cycle, Brazil contributed 21% of global sugar production and 26% of global ethanol production. Brazil's current standards require a 27% blend of ethanol in gasoline. Australian petrol stations sell a blend: 'E10', which is petrol with 10% ethanol.'

'Can ethanol be used in cars neat?' asked Biggs. 'I heard a story that when petrol was rationed in WWII, a tycoon ran his Rolls Royce on whiskey.

'Pure ethanol can be used as a fuel, but the high water content in whiskey makes it unsuitable without engine modification,' I said.

'Next reader, please.'

'Whiskey typically contains about 40% ethanol (alcohol) by volume, because distillation produces an azeotrope with 60% irreducible water. The remaining water would have to be removed chemically.

'Standard gasoline engines are not designed to run on alcohol fuels. Ethanol requires higher compression ratios and different fuel-air mixture ratios. Modifying an engine to run on whiskey would involve significant changes, such as adjusting the fuel injection system and possibly changing materials to handle the corrosive nature of alcohol. Even if you managed to modify the engine, whiskey would not provide the same energy output as gasoline. Ethanol has a lower energy density than gasoline, so the engine would produce less power and be less fuel-efficient.'

'Do they drink it?' asked Leblanc.
'Next reader, please.'

'Yes,' I said. 'Rum is made by fermentation of sugar. Energy can be released from plant material by enzymes, such as yeasts and by bacteria. Distillation follows in stages. Gasohol, a gasoline substitute, is a blend of petrol with alcohol from sugar cane in Brazil and from corn in the USA. It extends imported gasoline in Brazil. The USA and Brazil are the biggest producers of plant ethanol but latest reports are that as the environmental costs of producing them have become better known, they have seemed less attractive. Growth of bioenergy has faltered and attention is returning to conventional fuels and the other sources of renewable energy, solar, wind and hydro. It is more profitable to use the land for growing crops.'

'The environmental gains you get from growing plants, due to photosynthesis, are taken away when you ferment the plant material,' said Biggs. 'Alcohol production doesn't get you ahead by reducing carbon dioxide.'
'Even drinking it wouldn't be of much benefit,' said Kelly.
'A lot depends on your tolerance of carbon dioxide emission,' I said. 'In our next lesson we'll see if carbon dioxide is just a big scare.'
'You can't say that,' said Norman. 'People are too invested in the evils of carbon dioxide.'
'You are right, Norman. However, I believe in honest science and I'm not beholden to the doomsayers. I hope you will bring an open mind to our discussion.'
'It's phony to want ethanol because it's green,' said Tracey. 'Fermentation produces ethanol from cellulose, plant sugar, returning to the air the carbon dioxide absorbed by the cane when it grew.'
'Benefits of producing biofuels compared with renewable sources of energy are that the product is a liquid and easy to store. Disadvantages are that growing the plants takes valuable land area and water from other crops and its cultivation and harvesting consume energy.'
'Next reader, please.'

During the discussion. the students had been scrolling their phones.
'Ethanol takes large areas of good quality land and needs a lot of water,' said Kelly. 'Critics online believe that widespread production of ethanol will result in more land being used to grow corn for fuel, rather than for food. They also believe that producing and using ethanol actually does more harm than good to the environment.'
'They say ethanol and ethanol-gasoline mixtures burn cleaner and have higher octane levels than neat gasoline, but they also have

higher evaporative emissions from fuel tanks and dispensing equipment,' said Tracey, reading from her phone. 'These evaporative emissions contribute to the formation of harmful, ground-level ozone and smog.'

'Ethanol is likely to be at least 24% more carbon-intensive than gasoline due to emissions resulting from land use changes, from growing corn, along with processing and combustion,' said Kelly.

'What does 'carbon intensive' mean?'

'It could be something like the amount of carbon used per vehicle kilometre,' I said. 'It's a complex story and it has changed, becoming less attractive with time. Fortunes have been poured into ethanol projects and some may have been uneconomical and shut down. There has to be caution in selecting energy alternatives and preparedness of science to catch up.'

In the next chapter we will consider the consequences for the greenhouse effect of producing carbon dioxide and methane.

TASK 15
List advantages and disadvantages of ethanol compared with petroleum for cars.

TASK 15 Answers
Ethanol burns less easily in internal combustion engines.
With ethanol, the range of the vehicle is reduced.
Ethanol is obtained by growing plants, taking land out of other production.
Ethanol has different pollution effects.

CHAPTER 16 THE GREENHOUSE THEORY

'Earth's climate is kept warm by its unique atmosphere,' I said. 'The Sun heats the near-side of the Earth, which rotates into shadow, where heat is lost into the atmosphere and into space by radiation. The meaning of 'greenhouse effect' is the insulating effect of the atmosphere. 'Greenhouse gases' used to mean all the gases in the atmosphere before carbon dioxide was identified as a pollutant, whereas now 'greenhouse gases' commonly means carbon dioxide, at only 0.042%, with a little methane and sometimes up to 4% of water vapour, although the water vapour is usually not mentioned, nor are nitrogen and oxygen which make up most of the atmosphere and provide most of the Earth's insulation. The 'greenhouse' aspect was that the gases were recognised for holding in the heat plants needed for growth, as did glass in a 'greenhouse'.

'Biggs will you read, please.'

'When scientists observed that the Moon and Mars alternated from hot to very cold during their daily rotation, they wanted to explain how the Earth stayed at moderate temperatures. They attributed Earth's climate to its dense atmosphere, whereas the Moon and Mars had little or no atmospheres. Earth's atmosphere insulated it from solar heating and from cooling into cold space. Earth's temperature averaged between hot and cold, with the atmosphere smoothing the days' highs with night time lows. The atmosphere was envisaged as a layer of insulating gases more than 100 kilometres thick, which moderated temperatures into a warm range, that people are adapted to live in.'

'Do you believe it?' Tracey asked me.

'Yes, it seems reasonable,' I said. 'But there has been a complication. They have wanted to explain why Earth has seemed to be warming and CO_2 has captured attention, because its concentration is increasing steadily, at seven times faster than global temperature. A complex mechanism detailing the role of CO_2 was invented and named an Enhanced Greenhouse Effect (EGHE). It was an extension of the greenhouse effect, which had been regarded as benign, with additional CO_2 transforming it into a malignant problem of major global significance.'

'Next reader.'

'There have been various attempts to explain how 0.042% of the atmosphere causes the other 99.958% to warm disproportionately. The role assigned to carbon dioxide seems counterintuitive. I accept that the molecules could possibly trap more infrared than nitrogen and oxygen, but I dispute that the effect is large. Measurements of absorption by infrared and Raman spectroscopy bear little relation to absorption of solar energy at the Earth's surface. The effect of additional CO_2 has attracted an insane quantity of theoretical speculation and is still not clear.'

'CO_2 had been benign, but when it increased, it was labelled a pollutant?' said Pamela.

'The climate science articles I have read have various explanations of the EGHE. I found none of them convincing,' I said.

'Next reader, please.'

'The greenhouse allusion is confusing. In a real greenhouse there is warming by the sun penetrating the glass and the heat cannot convect away because the glass is a physical barrier.

When there is no glass barrier, radiation penetrates to the Earth's surface and the Earth is heated. The air convects away and heat is able to rereadiate into the atmosphere. They say that the 420 molecules in every million which is carbon dioxide, or one in 2381 molecules, traps enough heat to cause warming of the planet. This is counterintuitive.'

'It is,' said Jessica. 'Could such a dilute presence really trap enough heat to warm the whole atmosphere significantly?'
'I don't believe it could,' I said.
'Next reader, please.'

'The detailed explanation they give rests on the unique infrared absorption property of carbon dioxide molecules. Each molecule is supposed to transfer more heat to the atmosphere than all of the other 99.958% of gas molecules,' I said. 'The process that enables a very small quantity of carbon dioxide to absorb infrared at the surface and mingle with and heat up a very large mass of atmospheric gases has not been sufficiently explained.'

'Do you believe it?' asked Pamela
'No. I have not seen a satisfactory explanation,' I said. 'The usual scientific evidence has not been available because the scale of the system is too large and complex to model reliably.'
'Why don't you believe the infrared is absorbed by the atmosphere?' Pamela asked.
Next reader.
'They use spectroscopes to measure absorption and emission of light and other radiation, that do not simulate atmospheric conditions,' I said. 'You can feel how infrared warms air at pavement restaurants with infrared heaters. Their heat does not warm the air directly,' I said. 'Diners can feel the infrared warmth on their skin, but the air they breathe does not warm up until there has been convection and radiation from the hot surfaces which the infrared has reached.'

'Wow!' said Kelly. 'The atmosphere is heated mostly by the small amount of radiation reaching the Earth's surface and not reflected.'
'Next reader.'

'I believe infrared from the sun warms the Earth's atmosphere from absorption on the Earth's surface and convection into space, without carbon dioxide molecules having a disproportionate effect. All of the gases of the atmosphere help in insulating the Earth.

'So the increasing concentration of carbon dioxide is a coincidence?'
'Yes. It actually results from ocean warming. I doubt that it warms the rest of the atmosphere.'
'Earth warming has been overlooked from emission of combustion heat.'
'I thought heat transfer was simple,' said Emily. 'I didn't know science was this interesting,'
'Perhaps the small amount of pollution has a disproportionately large effect on the temperature of the atmosphere,' said Norman.
'I don't think so,' I said. 'CO_2 is not a known catalyst. There is a belief that global warming can be stopped by reducing certain trace gases, as was done with refrigerator gases causing the ozone hole. A small quantity of those are supposed to have a very large effect, supposedly breaking down the ozone molecules, without ever being used up. CO_2 does not have a chemical catalytic effect like that. To cast CO_2 as a catalyst is fantasy.'
'Perhaps CO_2 has some other effect that we don't know about yet?' asked Sophie.
'Some people believe in the Tooth Fairy,' said Buck. 'They live in hope.'
'Are you saying the infrared absorption propensity of carbon dioxide is a red herring?' asked Tracey.
'Yes. I'm saying that the number of carbon dioxide molecules absorbing the infrared is too few to warm the entire atmosphere.'
'So CO_2 does not cause significant heating of the atmosphere?'
'I think not.'
'Of course it doesn't,' said Norman. 'The Enhanced Greenhouse Effect is bull shit.'

'No-one is listening to you, Norman. Can you show the climate scientists are all wrong? Of course you can't.'

'They'll be found out when their theories become more and more complex,' Norman said. 'In the meantime I can cast doubt on their predictions. People shouldn't let them destroy technology that has served us for hundreds of years.'

TASK 16
What was the effect on the Greenhouse Effect of growth of carbon dioxide from 370ppm to 420 ppm on the Keeling Curve?

TASK 16 Answer
Global warming commenced. Climate change is said to have started.

CHAPTER 17 GENESIS

'Two different explanations of the origin of the Earth are in the Bible's Genesis and the Big Bang theory. The same method can be used to compare the movie's Greenhouse Theory with my sceptical science, to find out which to believe.'

'Do you mean the origin of things like washing machines?' asked Buck. 'The theory is they originate in factories.'
The others laughed.
'No, Buck. I mean things that aren't manufactured, like the ground we're on, the oceans, the air we breathe and all living things?'
'Oh,' said Buck, 'natural things.'
'Yes, how do they originate?' I asked.
'Nature,' quipped Buck and the others laughed. Tracey and Pamela thought his answers were hilarious. I found his opposition wearing. But his responses had the group's interest.
'Where does nature originate?' I asked.
'The birds and the bees,' he said. 'I heard that is where the action is.'
We all laughed.
'Nice try and partly right, Buck,' I said. 'But there's more to it than reproduction. In fact there are two very different stories about where everything has come from: Genesis in the Bible and The Big Bang Theory. We'll look at the Bible today and the Big Bang next lesson.
'We'll read Genesis now. Your job will be to decide which story you find more convincing.'

'The First Book of Moses.
GENESIS

1 In the beginning God created the heaven and the earth.
2 And the earth was without form, and void, darkness was on the face of the deep. And the spirit of God moved upon the face of the waters.
3 And God said, Let there be light: and there was light.
4 And God saw the light, that it was good and God divided the light from the darkness.
5 And God called the light Day, and the darkness he called Night. And the evening and the morning were the first day.
6 And God said: Let there be a firmament in the midst of the waters, and let it divide the waters from the waters.
7 And God made the firmament, and divided the waters which were under the firmament from the waters which were above the firmament: and it was so.
8 And God called the firmament Heaven. And the evening and the morning were the second day.
9 And God said, Let the waters under the heaven be gathered together unto one place, and let the dry land appear: and it was so.
10 And God called the dry land Earth; and the gathering together of the waters called he Seas: and God saw that it was good.

So far about half the class had slumped forward across their desks, or laid back in a sleeping posture.

11 And God said, Let the earth bring forth grass, the herb yielding seed, and the fruit tree yielding fruit after his kind, whose seed is in itself, upon the earth; and it was so.
12 And the earth brought forth grass, and herb yielding seed after his kind, and the tree yielding fruit, whose need was in itself, after his kind: and God saw that it was good.
13 And the evening and the morning were the third day.
14 And God said, Let there be lights in the firmament of the heaven to divide the day from the night, and let them be for signs, and for seasons, and for days, and years.
15 And let them be for lights in the firmament of the heaven to give light upon the earth; and it was so.

16 And God made two great lights; the greater light to rule the day, and the lesser light to rule the night; he made the stars also.
17 And God set them in the firmament of the heaven to give light upon the Earth.
18 And to rule over the day and over the night, and to divide the earth from the darkness; and God saw that it was good.
The students were inert except for yawning and moving into more comfortable positions.
19 And the evening and the morning were the fourth day.
20 And God said, Let the waters bring forth abundantly the moving creatures that hath life, and fowl that may fly above the earth in the open firmament of heaven.
21 And God created great whales, and every living creature that moveth which the waters brought forth abundantly; after their kind, and every winged fowl after his kind and God saw that it was good.
22 And God blessed them, saying, Be fruitful, and multiply, and fill the waters in the seas, and let fowl multiply in the earth.
23 And the evening and the morning were the fifth day.
24 And God said, Let the earth bring forth the living creature after his kind, cattle and creeping thing, and beast of the earth after his kind: and it was so.
25 And God made the beast of the earth after his kind, and cattle after their kind, and everything that creepeth upon the earth after his kind: and God saw that it was good.

'Is there much more of this, Sir,' asked Kelly, yawning.
'The Bible is a thick book,' I said. 'Not much more today.'

26 And God said let us make man in our image, after our likeness: and let them have dominion over the fish of the sea, and over the fowl of the air, and over the cattle, and over all the earth, and over every creeping thing that creepeth upon the earth.
27 So God created man in his own image, in the image of God created he him, male and female created he them.

28 And God blessed them, and God said unto them, Be fruitful, and multiply, and replenish the earth, and subdue it: and have dominion over the fish of the sea, and over the fowl of the air, and over every living thing that moveth upon the earth.
29 And God said Behold, I have given you every herb bearing seed. Which is upon the face of all the earth, and every tree, in the which is the fruit of a tree yielding seed, to you it shall be for meat.
30 And so every beast of the earth, and in every fowl of the air, and to everything that creepeth upon the earth, wherein there is life, I have given every green herb for meat: and it was so.
31 And God saw everything that he had made, and behold, it was very good. And the evening and the morning were the sixth day.

'I'll stop there,' I said.
'Thank God for that,' said Buck. 'I'm creeped out.'

The other students were waking up. My religion topic had created an apathetic torpor. My lesson was supposed to contrast two different theories but the students had no interest so far.
'Genesis commences the Christian Bible,' I said. 'It was first written by Moses and its final version as late as post-exilic Israel around 400 BCE.'

The students were unimpressed. Their apathy was forgivable. It was irrelevant in their worlds. I tried to wake them up with a summary.
'This first verse, began with the earth barren, without living things,' I said. 'The spirit of God created light and heaven and divided the waters in seas to form Earth as dry lands with seeds to grow grass and fruit. He divided day and night with lights. Thus the heavens and the earth were finished, and all the host of them.'
The students were awake now, yawning and eye-rolling. I tried to get them into their task of comparing the Bible and Science.

'Whereas Science has a story about exactly what happened after the Big Bang, the Bible's account of how the Earth began is difficult to take literally.

'The Bible doesn't explain where God came from and why he had taken on organising light, land and living things, large and small. He set them to growing and reproducing. He formed man of the dust of the ground to till the ground and eat of every tree, except the tree of knowledge of good and evil. The first man was Adam and when he was asleep the Lord God took one of his ribs and made woman.

'The two accounts have several differences, such as explanation of the origin of men and women.'

'Cool,' said Kelly, waking up. 'Did Adam take a rib, like a gardener taking a cutting and planting it? Or did he graft it on to a female?'

'Then the woman would have the same cells as Adam,' said Tracey.

'Then why are men and women different?' asked Jessica.

'They aren't different,' said Kelly sardonically. 'It's an illusion kept going by the fashion and cosmetics companies to sell products.'

They were all awake now.

'But a woman can have a baby,' said Jessica. 'Sorry Kelly, but men missed out.'

'It's not missing out to forego an unpleasant experience. '

'I'm glad you willingly accept the male role is accessory, Kelly,' said Jessica, smirking. 'Parenting would be too unpleasant for you.'

'That's enough,' I said. 'This discussion won't resolve anything. It seems that the biblical explanation of the origin of dimorphism is controversial. Is the Big Bang theory more agreeable?'

'What is it?' asked Norman.

'Read the next chapter for our next lesson,' I said.

TASK 17
Answer this question.
List the main events listed in Genesis when the earth was formed.

TASK 17 Answer
Darkness, Light, Night, Day, Waters, Firmament, Dry Land, Earth, Seas, Grass, Seed, Fruit tree, Seasons, Days, Years. Creatures, Fowls, Multiplying, Creeping thing, Man, Male, Female.

CHAPTER 18 BIG BANG

'In this lesson we'll compare the Big Bang Theory with the Genesis description of the origin of the Earth that we read from the Bible last time. My Big Bang Theory is bits and pieces retained in my memory, of events I have heard from scientists.

'The Big Bang occurred at a point in empty space where all matter was concentrated. For some reason, the mass exploded, propelling debris outwards. It is still moving outwards today, to a distance of 14.5 billion light years.'

'What's scientific about that?' asked Norman.

'Observations of the position and sizes of objects in space agree with the theory. Space has various celestial objects. The explosion scattered clumps of matter which are held together by gravitational forces. There are millions of vast galaxies and their star systems, with suns, planets, comets and mysterious dark matter. Earth is a planet of a middle-sized Sun on an arm of the Milky Way galaxy.

'4.5 billions of years ago Earth was a clump of material formed from the gravitational attraction of debris from the big bang itself or from supernovae when there were further bangs. Slowly the mass collected and heated up becoming a very hot sphere, the Sun. It threw out material that cooled and formed the Earth and planets, solidifying on the outside to form plates of land floating on magma. Moisture in the atmosphere cooled and oceans formed. The first life was possibly microscopic, something like particles of polymeric clay-like material. They evolved into microbial animals that reproduced and grew in the phyla to become mammals, fishes and birds. Modern man evolved from an ape ancestor, a humanoid whose modern descendants have transformed the planet and are perceived to threaten its future.

'It's a tall story,' said Sophie, 'with more imagination than measurement.'
'Is there any record of animals changing into mammals, fishes and birds,' said Norman. 'Is it possible God made them from a dinosaur's ribs, or some goddam thing?'
'It's possible, but it is easier to explain many small changes in a long series.'
'There have to be step changes too,' said Norman. 'For example, it's difficult to imagine what existed before the very first eye appeared, a zebra's stripes, and the peacock's tail.'
'When there are large steps, perhaps intermediate steps are not yet known,' I said. 'Evidence for natural selection is in some instances deduced and may be without the definite causality required to verify the theory. Is the Big Bang Theory at all credible?
'I like the Big Bang Theory because it has hypotheses with processes that continue today and can explain observations of the earth and living things all around us,' said Tracey. 'Even so, much of the evolution theory is conjecture. For example, although the Big Bang Theory is supported by evidence of the outward movement of material from a point source, we know little about the mass that exploded, how it got there and where it came from.'
'Well said, Tracey,' I replied. 'It is difficult to assert the reality of the Big Bang Theory when it has so many holes.'
Next reader please.

'The biblical account has a detailed timetable, unbelievably vast and quick for the enormity of events, which are vague but have plausible processes. The narrative of Genesis is a diary of God's acts in transforming the Universe from a featureless barren place in darkness into seas and earth growing grass and fruit in daylight, with animals, birds, insects, a man and a woman all miraculously created.

God's project took him only 6 lousy days, which is preposterous from a science viewpoint.'

'In the Big Bang Theory,' said Kelly, the origin of God and the transformation of the Universe from the state God inherited is left to the reader to imagine, with the dots joined by familiar processes, taking millions of years, rather than having definite series of intermediate forms. Neither Genesis nor the Big Bang have dates for when Earth and the Universe originated. Apart from the date of the Big Bang, the science theory has many hypotheses about how the Earth could have formed. The Genesis account does not explain how the Earth has come to be the way it is. The role of the text acknowledges that God was instrumental in providing original features of the Earth but not much after that.'

'Yay, Kelly,' said Tracey. 'The Big Bang has it.'

I wrapped the lesson up.

'The pretexts of both accounts of the origins of the Earth are unsatisfactory. They are reduced to religious dogma in Genesis and to pseudoscience in the Big Bang Theory. Genesis is a homily, a spiritual acknowledgement of God's power,' I said. 'We are supposed to be overawed. The Big Bang Theory better satisfies our curiosity about the relationships of celestial objects to each other and the evolution of living things. It incorporates the theories of scientists like Copernicus, who in the 16th Century revealed that the Earth was a planet orbiting around the Sun.

'We have compared two different theories and opinions, to find which is better, but neither is satisfactory,' I said. 'Next lesson we'll compare two theories about climate change to evaluate them. Comparison is controversial and it may not be possible for you to remain as uncommitted as you have been in considering these two theories.

'Put your arm up if you prefer Genesis?' I said.
One arm, Tracey's, was raised.
'Put your arm up if you prefer the Big Bang?'

About half the students put their arms up.

'I assume the rest of you are 'Don't Know?' or prefer an alternative?'

'We don't give a fuck,' said Buck.

Everyone laughed.

I couldn't argue with them. The lesson ended.

Tracey was the last to stand up.

'Thank you sir. That was a great lesson.'

'What did you like most?'

'It was the first time I have heard the Bible discussed in a science class. Before, when God was mentioned, it was always derogatory. I believe in God and the Bible and today I felt respected. I know that you are not a believer, but you don't impose your views. My view has been counted. I feel about ten feet tall.'

TASK 18

List observable features of the early Earth according to the Big Bang theory.

TASK 18 Answers.

Explosion, Dispersal radially, Agglomeration of debris, galaxies, suns, molten Earth, planets, moons, seas.

CHAPTER 19 EMOTIONAL TRUTH

'In this lesson we'll use the same method we used to compare Genesis and the Big Bang to make a comparison of two theories about climate change. You'll watch Al Gore's emotional climate movie, 'An Inconvenient Truth', 2006, compared with my own sceptical science version of climate change. I will explicate my views in the following pages.

'Gore is a US politician and his invective is like a documentary with graphs espousing dramatic climate relationships. When I first saw his movie, I was convinced that the planet was being damaged and something had to be done urgently. Now I've changed my mind and although I still accept that damage could be concerning, there is no crisis and the amount of warming is not alarming in my opinion. I'll be interested to hear your views. I'll ask you to choose between the two theories when we've discussed them.

'Why did you change your mind about the movie?' asked Sophie.
'I watched other movies that were made in response. I investigated but couldn't figure out how his theory about pollution and the Greenhouse Effect was different from my understanding from my experience of heat transfer in gases and radiation. His theory seems wrong. Carbon dioxide is not a threat, in my opinion. I will explain my views in the coming lessons.'
'Why is your viewpoint so unpopular?' she asked.
'You should ask those who believe the movie,' I said. 'I suppose they think because there are more of them I should believe it too.'
'It is widely believed by scientists,' said Tracey.
'The Phlogiston was widely believed to be a substance involved in combustion from 1667, until Priestly and Lavoisier named oxygen in

1777. Most scientists were wrong for a century,' I said. 'Have any of you seen the movie before?' I asked.

Three arms were raised.

'When you watch it, remember any parts that could be false, for discussion afterwards.'

It was a long video and we watched it straight through. All the class seemed alert but I wasn't sure how much they were taking in. They were in a sombre mood as the movie finished.

'There was nothing false, Sir,' said Norman. 'I believe the video is absolutely true.'

'It was exaggerated,' said Tracey.

'It was bull shit,' said Biggs.

'We'll discuss the main points,' I said. 'I have summarised in this chapter the arguments in the video, in order. Would you start reading my notes.'

'First reader, please.'

'Gore quotes environmentalist Carl Sagan as saying that the Earth's atmosphere is thin enough that we are capable of changing its composition and states that pollution is causing global warming by thickening the atmosphere,' read Kelly.

'Carl Sagan is normally reliable,' I said, 'but I believe he has exaggerated the fragility of the planet. The atmosphere begins at a distance of 6371 kilometres from the centre of the Earth, mostly in a layer 100 kilometres thick, with a mass of 5.1 million billion tonnes. This mass is 80,000 times more in mass than the 63 billion tonnes per annum of carbon dioxide produced by fossil fuels consumption. To double the concentration carbon dioxide would take 81,000 years if all the CO_2 emissions were retained. But a large part dissolves in the oceans or escapes into space. Carbon dioxide is a mere trace pollutant, highly dispersed, rather than the massive intrusion suggested by the movie.'

'There doesn't seem to be much carbon dioxide given all the fuss there is about it,' said Jessica. 'Even if it was red hot, it wouldn't heat the air much.'

'It took only a little chlorofluorocarbons (CFCs) to close the ozone hole, because they were a catalyst,' said Tracey. 'If carbon dioxide was a catalyst, a little could cause a reaction and it wouldn't be used up.'

'That's correct and a good point. However, carbon dioxide does not have any known catalytic effect with nitrogen, oxygen and the other gases,' I said. 'To have such a big effect when so dilute, the effect would be extraordinary and unknown to science.'

'Is that possible?' asked Norman.

'Yes,' I replied. 'The chemistry of carbon dioxide is well known. But it is unlikely there is undiscovered catalysis.'

'They couldn't figure out how green plants grew until they discovered carbon dioxide and the catalyst chlorophyll. Science mostly makes progress slowly. The effects of trace elements can be surprising. An example is how the Ozone Hole forms and disappears on an annual basis, in springtime over Antarctica. They thought that they had nailed it by banning CFCs, but the hole keeps coming back.'

'Carbon dioxide's heating effect is doubtful,' Tracey said. 'It's presence maybe be less consequential than the CFCs were supposed to be.'

'I agree,' I said. 'The movie didn't describe the atmosphere accurately. It adheres to the Earth because its molecules revolve around the Earth with insufficient kinetic energy to escape the inward pull of gravity, which stops them from being thrown outwards into space. A few gas particles are lost and others captured, maintaining atmospheric pressure. It seems unlikely that the atmosphere could be thickened significantly for long, although the concentration of one particle type could slowly increase if released at a high enough rate from sources on the Earth's surface.'

'The movie said the carbon dioxide builds up,' said Kelly. 'Why would it, if it is being thrown outward?'

'Good point, Kelly. Some of it would dissolve in the oceans, perhaps most of what comes from power stations. The rest will build up until it reaches equilibrium when the same quantity is being lost into space.'

'The movie showed carbon dioxide concentration increasing fast,' said Jessica.

I projected the graph below onto the screen.

'There is a trend of steady increase in CO_2 concentration which looks fast because the concentration moved up to 310 ppm at the bottom corner, multiplying the steepness by 4. The increase is really quite modest. Also, the curve has a saw tooth appearance, which could possibly be biannual rather than the indicated monthly cycling.'

'The curve is important evidence, is it not?' asked Leblanc. 'It should be clear what it shows.'

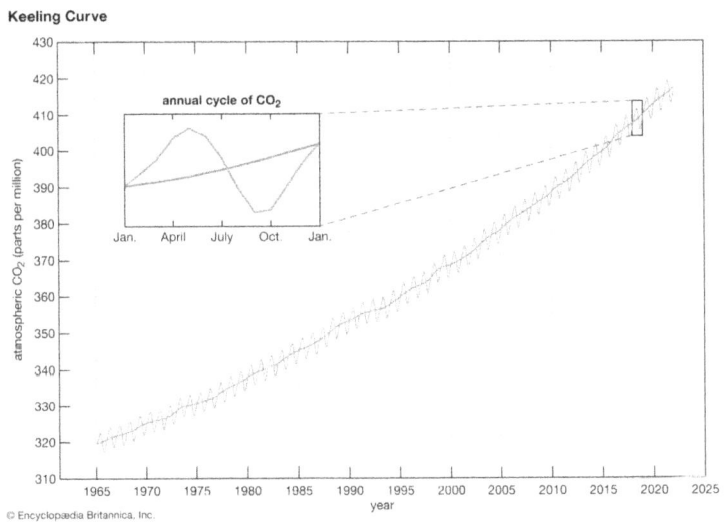

Source: https://www.britannica.com/science/Keeling-Curve

'Yes, that graph from Hawaii, the Keeling Curve, has attracted a lot of attention,' I said.

'What does the graph mean?' asked Jessica.

'It shows carbon dioxide concentration in the air is increasing with time,' I said. 'The suggestion is it will continue while existing conditions continue. But I attribute most of the increase in carbon dioxide to ocean warming. Carbon dioxide bubbles out of the sea, like the fizz from a warming glass of cold lemonade. Oceans contain large quantities of dissolved carbon dioxide that would be released with warming. The ocean is slowly warming and the solubility of carbon dioxide slowly falling due to thermal emissions. This could explain from where up to 100% of the measured increase of CO_2 in the atmosphere came from. I believe the causality is the reverse of the movie's suggestion: the CO_2 is released by warming of the ocean; the CO_2 does nothing to warm the Earth or the oceans it dissolves in.

'What a lousy trick,' said Kelly. 'The bastard movie fooled me.'
'It's a bloody fraud,' said Jessica.

TASK 19
Why is atmospheric carbon dioxide increasing?

TASK 19 Answers
Most of the Earth's carbon dioxide is dissolved in the oceans. Atmospheric CO_2 is in equilibrium with dissolved CO_2 regulated by known partition coefficients. When the ocean temperature increases, CO_2 is released and the atmospheric concentration of CO_2 increases.

CHAPTER 20 CAUSE AND EFFECT

'The increase in CO_2 was by dissolution from the ocean caused by warming, not warming caused by CO_2 entrapment. It was the reverse of conventional thinking. It was a better explanation, because by Ockham's Razor, it was simpler.

'The graph showing increasing CO_2 in the air could be evidence of ocean warming without portending disastrous global warming. The amount of CO_2 is an effect, not a cause, of climate warming.

The movie has falsely asserted a strong link of human industrial activity, when CO_2 could result from several different causes, especially ocean warming, as we have seen above. Another cause could be the increase in water vapour in the atmosphere. It too is an emission and would add more to temperature increases than carbon dioxide.

'So carbon dioxide results from warming, not vice versa,' said Leblanc.

'Yes. Climate science has inverted the causality.'

Several students laughed nervously, worried that I was disrespecting a movie which was a foundation of the climate warriors' beliefs.

'Should you be criticising the graph when so many people believe it?'

'Yes, absolutely. Science isn't done by agreeing with powerful groups. The movie's claims contradict physics and chemistry theories that scientists have trusted for hundreds of years. Where the movie's claims don't fit the theories, I believe it is my duty to call them false.

'There are other problems with Gore's information,' I said. 'He supposed that ice cores can be recovered intact, preserving air frozen in glacier water 650,000 years previously and measuring the quantity of CO_2.

'What's the problem?' asked Norman.

'Can you imagine drilling ice cores buried in an ice sheet for 650,000 years and recovering dissolved gases, keeping historic pressures on the core samples, collecting the tiny amount of carbon dioxide as the core melts and the water warms up?' I said. 'Could the gas accumulate, leak or escape during drilling, or during gas recovery, or during analysis? If they told me carbon dioxide in the ancient air was 420 ppm, because that is what they measured, I would say bullshit. I would want to know the standard error of the measurement method because it would be a wide range that would cast doubt on the graph of increasing CO_2 above.

'They wouldn't have used any old method,' said Tracey. 'Scientists usually believe in their data.'

'That's true,' I said. 'They had no good way of checking their findings. They could have been desperate to make a measurement and leave it to someone else to question it, as we are.'

'Ice core measurements are disputed by Ian Plimer, a well known geologist,' I said 'Taking a wide view of the history of Earth, he proposes that carbon dioxide has on several occasions been much more concentrated than it is at present and that the concentration is now relatively low. He explains the slowly increasing temperature to be due to continuing temperature cycles, or ice ages.'

'Do you believe that carbon dioxide is not a problem?' asked Biggs.

'That is my conclusion,' I said.

'Too bad that a large amount of money has been wasted,' said Kelly.

'The amount of the carbon dioxide pollution is tiny. Gore plays the emotions card in appealing to preserve current conditions for our children's futures,' I said. 'He presents a graph with variations in the

Earth's temperature, as if that could be recorded continuously by a single instrument. The amount of warming is under 0.5°C and the recording station details of location and statistical analysis is not given.

'The movie data was challenged, in an Oregon courtroom, about authenticity of a recent uptick in a hockey-stick graph, in a defamation trial lasting many years. When the rigours of aging prevented Professor Tim Box attending the court for the defendants, the defamation was dismissed and costs awarded against the litigants. Graphs displaying a dramatic 'uptick', such as Gore's graph, are no longer used in reliable publications.'

'The deception was very bad,' said Kelly. 'It's corrupt and shows the movie can't be trusted.'

'Gore's argument reverts to attributing bad weather, hurricanes and storms to carbon dioxide's climate effect, as if they could cause all bad weather,' said Tracey. 'He describes mechanisms by which global warming could result in disasters. It is unfounded speculation.'

'Absolutely,' I said. 'Since 2006 many of his predictions have fallen short.'

'No number of photos of bad weather can indicate climate change,' said Norman. 'He must assume viewers are stupid. That is insulting.'

'I saw on the news that the ice caps on Mars have melted,' said Jessica. 'It would not be due to atmospheric warming because Mars has no atmosphere. The scientists said it could be from sublimation into space, due to a change in the Sun, or in the orbits.'

'Thank you Jessica. Quite right. We don't have to believe the movie's idea that polar melting is due to global warming.'

'Gore criticises scientists who disagree with his theories,' I said. 'He says they have an independent obligation to respect and present the truth as they see it. That seems to be a contradiction, because science has stopped talking about truth and consensus and Gore's opponents can't get a hearing. He accuses them of misconceptions, but deceives us that science is done by consensus. It is not. Scientists have an obligation to make their findings available to other scientists and communicate independently, resisting others who attempt to drive them in a herd. Gore is a hypocrite.'

'Gore is not a scientist and his theories are not able to be falsified,' I said. 'He has tried to frighten scientists into staying silent. The worst is that he has too often succeeded. He invokes post modern philosophy, that does not rely on empirical evidence. He has used a wide variety of information sources to support his claims and they can be contested.

'Gore's paper is political – an invective promoted by rhetoric with appeal to his celebrity status and political connections. He is not a scientist. His university research thesis is reported to have proposed that political change should present big lies, rather than small ones, to affect audiences emotionally, rather than cognitively, with more effect. That is what he has done: told big lies with emotion to persuade people. The movie 'an inconvenient truth' was a moral crusade and did not consider poor nations who cannot afford expensive energy, nor developing nations, who do not have alternative supplies.'

'So are you saying Gore supported lies with emotion?' said Tracey. 'Is that how the post-truth philosophy works?'

'It is what I am saying,' I said, 'and Gore's theory is post-truth. It is the method of persuasion common in politics. It is not respectable in the guise of science. 'An inconvenient truth' has several big lies and is steeped in emotion and politics. It omits to address matters of obvious importance to Gore's thesis that should have been included, for example how the small amount of carbon dioxide could heat the atmosphere. The implications for developing countries and people living in poverty are dire.'

'I don't regard this movie to be an honest communication,' Tracey said.

'There have been several refutations by distinguished scientists. Perhaps it will be discredited by courageous commentators soon.'

TASK 20

Recall complaints we have with this movie.

TASK 20 Answers

The trapping of heat by carbon dioxide is not believable.

The warming of the total atmosphere by infrared absorption by a tiny amount of carbon dioxide is not possible.

The temperature uptick was fraudulent.

The air recovered in ice cores is not to be trusted for authentic historic levels of carbon dioxide in the atmosphere.

CHAPTER 21 EARTH'S CLIMATE

The aim in this chapter is to reveal some of Earth's climate processes, elucidating processes misrepresented in the movie 'An Inconvenient Truth.'

'The movie said the tiny amount of radiation trapped by carbon dioxide could heat all of Earth's atmosphere,' Jessica complained. *'But it didn't explain how?'*
'Good question Jessica,' I said. *'I think we can assume that no such heating method exists beyond the ordinary Greenhouse Effect that we studied.'*
'The movie pretended that scientists were all in agreement, when they weren't,' said Norman. *'The movie's view is narrow, the opinion of only a few people, not necessarily well qualified.'*
'I agree, Norman,' I said. *'I dispute that carbon dioxide is the only emission from combusting fossil fuels that causes the small amount of global warming. Heat radiation and water vapour cause warming too.*
'Next reader, please.'

'My theory does not have a movie, nor a politician and my ideas are supported by logic and simplicity, not by photos of extreme weather. The movie does not explain what happens to the Sun's energy when it is radiated to Earth. I will begin from the outside, considering the energy that comes to Earth from the Sun as radiation.'
*'The Sun warms the Earth as it rotates daily and revolves annually around the Sun. Solar radiation penetrates the Earth's blanket of gases and is absorbed at the land and ocean surfaces, which radiate heat and warm the atmosphere by radiation, conduction and convection and at night allow the shadowed side to cool. Cooling is

delayed by the insulating effect of atmospheric gases, called the greenhouse effect. The amount of carbon dioxide in the atmosphere is small and slowly increasing. I would like to know why the effect of the increase in carbon dioxide is an 'enhanced greenhouse effect' and has been distinguished from the larger warming effect of CO_2 previously known simply as the Greenhouse Effect.'

'Is Earth's temperature prevented from cooling by all the gases?' asked Norman.
'Yes.'
'Next reader, please.'

'The gases blanket the Earth, insulating it from heating and cooling. It is said that carbon dioxide causes more warming than the other gases, but there is very little of it, only 0.042% or 420 molecules in one million.

'Attributing the small alteration in the delicate balance of Earth and Sun radiation to a byproduct of fossil fuel combustion is suppositious,' I said. 'Carbon dioxide was much more concentrated earlier in Earth's history. It was not calamitous in the past and any human effects on climate since the industrial revolution could be coincidental.'

'Is there something about carbon dioxide that holds heat in?' asked Tracey.
'Some climate scientists have claimed that CO_2 is different to the other 'Greenhouse Gases' in absorbing more infrared, but the significance of this is abstruse.'
Next reader.

'The concentration of CO_2 molecules in the atmosphere is so dilute, the effect is little. It has an absorption frequency window that catches some infrared, but it isn't hugely more than is caught by molecules of other gases,' I said. 'Carbon dioxide gas molecules absorb somewhat more infrared radiation apiece than nitrogen and oxygen molecules, but there are few molecules of the gas to trap it and convey energy on to warm the Earth. To suggest that the few molecules of carbon dioxide would do much of the heavy lifting, in warming and cooling the Earth, is ridiculous.'

'What does the infrared do?' asked Mark. He was a quiet student who asked great questions.
'Climate scientists say the CO_2 molecules that absorb IR heat the atmosphere but apart from a few collisions with other gases, the few heated CO_2 molecules that reach the Earth's surface would heat it and the other gases by convection.
Next reader.

'Most infrared passes through the atmosphere and is reflected into space or warms the Earth's surface. That is where the infrared heats the world. Gas molecules of all types collide with Earth's surface and many are absorbed,' I said. 'It is incorrect that carbon dioxide warms the air mass significantly from within by trapping infrared, except by convection like the other gases. The air mass gets colder with height and is heated from below at the Earth's surface. Some of 420 molecules in 1,000,000 in the atmosphere are energised by infrared at the earth's surface, releasing energy when absorbed, causing the atmosphere to warm up by convection and re-radiation.

'Are you saying that the infrared absorbed by a few carbon dioxide molecules warms the atmosphere hardly at all, but they warm the Earth's surface if they reach it and collide with it,' said Leblanc.
'Yes, that's what I believe,' I said.
'Next reader, please'.

'Carbon dioxide has been vilified in the movie because its increase in the atmosphere has happened to be when humankind has been combusting carbonaceous materials. There has been some global warming but associating it with increased carbon dioxide is weak and false.

'The premise of the Net Zero campaign is false, that further carbon dioxide emissions must be balanced by reduction in carbon dioxide emissions elsewhere to prevent climate change. Holding down carbon dioxide emissions won't make any difference to global warming, because it's a red herring.

'Coal is about 90% carbon and combusts to form carbon dioxide. The slow increase in carbon dioxide is sufficient, it is claimed, to cause significant global warming, although the process is obscure. For this reason alone, the Australian government is moving to replace coal fired power stations with renewable energy, with a significantly higher cost of electricity. It is a mistake.'

'If carbon dioxide is not warming Earth's climate, then what is?' asked Kelly.

'The amount of warming is small, about $1°C$ over the 150 years since 1880, I said. 'I believe most of the increase is entropy, which is energy at too low a temperature to use. All electricity and gas supplied to homes and also from cars is eventually becomes low temperature heat in the atmosphere and oceans. The warmest places on the Earth's surface convect their heat into the atmosphere, or absorb it into cold oceans, or radiate it away into space.

'Where does the entropy end up?' asked Biggs.

'Being useless, spent energy, it just hangs around, as low temperature heat, being slowly radiated into space,' I said. 'It has near-ambient temperatures and particles in chaos. Most of the waste heat from homes probably radiates into space within a day or two, or stays as global warming. When cloud cover prevents radiation, the vicinity may be warmed a little, by reflected infrared. On cold clear nights the warmth in air and water is radiated out to the low temperatures in space.'

'So we can't ever stop global warming until we stop using energy?' said Jessica.

'Exactly,' I said. 'We have watched a movie falsely claiming that global warming results from carbon dioxide. We have also seen there are several other possible causes of global warming. I would like you to choose which of these two approaches is more credible.

Raise your arm if you believe 'An Inconvenient Truth' is more credible than the alternative theories I have claimed'.

One arm was raised, Sophie's.

'Now who believes the movie An Inconvenient Truth is less likely?' I asked.

Most of the others' arms went up.

'Buck's arm stayed down, because he was usually disagreeable. Pamela didn't like to oppose authority.

'Thank you for your honesty,' I said. 'It could help to make the world a better place.'

TASK 21

How does climate science explain carbon dioxide's higher infrared absorption effect?

TASK 21 Answer

When a few of the 420 molecules of CO_2 in every 1,000,000 molecules of atmosphere are energised by infrared and are absorbed by the earth's surface, where they release energy, causing the atmosphere to warm up by convection and re-radiation.

CHAPTER 22 NUCLEAR ENERGY

'Nuclear power stations are weather-free alternatives to renewable energy for generating baseload electricity. They are being considered for installation in Australia.

'This chapter is about how nukes work, as a reading for a self-study task.

'You are to read the chapter together in-class and discuss it with others, for a test in two weeks' time.

Nuclear power is in the news and controversial. Thinking critically about it could be a new experience for you. The skills are logical reasoning, problem solving, critical evaluation and synthesising.

'The content you learn could be of permanent value, because nuclear energy could become a part of your lives, as it has been in mine. Your results will be part of your term's assessment, counting half as much as the exam. If there is anything in the text you don't understand, ask in class any day.

'The test questions are at the end. We will read through the chapter carefully, commenting critically.'

'First reader please.'

'When American scientists designed the first atomic bomb in 1945, it seemed possible it could unleash an explosive force like the reputed Big Bang and destroy the Universe. Explosion of two bombs killed about 250,000 Japanese, causing them to surrender. Thereafter the Allies developed nuclear fission as a slow 'explosion' able to generate electricity in the USA, UK, Germany, Canada, France, Japan and many other countries.

'My experience with nuclear energy follows, beginning in 1967.

'Hinkley Point Power Station in the UK was constructed on land taken from my family's farm. I continued to live there beside the power station until I left to go to university. My first job was there, as a vacation student. Part of my work was to test for radioactive leakages. I wore a film badge on my lapel, collected each week and checked for exposure to radiation. With excessive exposure, I would have been assigned to a less radioactive workstation. Workers had to familiarise themselves with different jobs. Nuclear power stations are dangerous workplaces. Dangers in coal mining could also be of concern for power stations, but nuclear stations have more insidious dangers.

'The reactor was a room-size block of graphite called an atomic pile, inside a pressured sphere with thick metal walls, having openings through which carbon dioxide gas at high pressure was circulated in and out. The hot gas coolant flowed to large heat exchangers where steam was generated for the turbines. The rest of the power station was steam boilers with turbo-generators, like in any coal-fired station. But the heat to generate the steam came not from burning coal, but from a radioactive chain reaction, slow to start and slow to stop.

'At Chernobyl there was an accident in 1986 and the pile began to meltdown through the concrete containment, into the rocks below.'

'Next reader please.'

'I was impressed that a year's supply of uranium fuel arrived in a lead-lined box on a low-loader with 120 wheels. The waste was taken away later in a lead 'coffin', to a plant at Sellafield, Europe's most hazardous nuclear site, for reprocessing. The flask contained about 40 kilogrammes of highly radioactive waste in a container designed to prevent rupturing, in accidental collision with a train, or from any action that could break it open, because it contained lethal plutonium.

'When fresh fuel rods arrived at Hinkley Point, they were stored temporarily in the 'swimming pool', a bath of caustic soda, where fissile rods from the core were brought to 'cool' off. The rods were cut out of their cans and welded into new ones, by robots working machine tools below the surface of the 'swimming' pool, controlled by people who could watch them work. At Fukushima in Japan a similar pool was submerged by a tsunami and leaked radioactive material.

'Next reader please.'

'That reactor core and other components wore out after 30 years and the massive reactor building, over 10 stories and the size of a soccer field, had to be cocooned in thick concrete poured from helicopters, because it couldn't be safely dismantled and replaced. Nuclear power station sites are lost forever, with no prospect of reclamation. The reactors are expensive to rebuild and could be taken over by terrorists. The equipment is too radioactive to safely remove.'

'This place on the planet would be removed from human occupation forever. The birds and wild animals that had lived on our farm at Bradland's Copse, before it was built, would never return.

'Although a nuclear power station neighboured our farm, we didn't experience problems. I expected that a third reactor would be built, but histrionics after the Chernobyl accident put paid to that. I felt that dangers from radiation and fall-out had been exaggerated. My view has been vindicated by recent reports from the Chernobyl wildlife reserve, where wolves, European bison and other species are thriving.

'Next reader, please.'

'The redeeming feature of nuclear power stations is reliable production of bulk cheap electricity. Although small reactors are possible, it is likely that large ones will be demanded for economy, centralising energy supply in stations over 1000 Megawatts, like it was in the large coal-fired stations spurned by the public.

'In developed countries electricity demand has run out of cheap options for supply. It does not seem reasonable to leave electricity supply in private hands when the escape of radioactive particles and costs of nuclear power would be borne by the community. Governments should intercede and prohibit nuclear energy because of the harm done to communities. If voters knew of the true dangers of nuclear power stations, it is doubtful new ones would ever be approved by a democracy.

'Interest in nuclear power has recognised their base load role would complement weather dependent and time-of-day dependent renewable energy. Solar panels have been able to make inroads into supply peaks during daylight hours. Battery storage has widened the role of solar panels but nuclear could takeover some of the heavy lifting previously done by coal.

'Nuclear power has a growing dilettante following. Construction is slow and expensive. Mining uranium and enriching it into fuel rods could face security and waste disposal difficulties at each stage. The technology can seem promising at a distance. Before accepting a proposal, the decision makers should visit a nuclear power station to realize that working conditions have dangers, security threats and unknowns.

'So in your view we shouldn't have nuclear power?' asked Leblanc,
'No, absolutely not. If we can't get electricity any other way, it would be better to go without it.'
'Rationing?' he said.
'Yes,' I said. 'Even log fires and wearing animal skins would be better.'

TASK 22
Answer these in-class test questions and hand them in at the next lesson. A wide range of responses is possible and will be evaluated for the four critical thought processes.
1. Logical reasoning.
   What were probable reasons for choosing carbon dioxide to carry heat out from the core?
2. Problem Solving.
   Would a nuclear power station be appropriate in Australia?
3. Critical Evaluation.

Compare wastes from nuclear and coal power stations, assuming a coal station of the same size could use 2 million tonnes of coal per year, with composition 10% ash and 90% carbon.
4. Synthesising.
Would film badges provide sufficient protection of the station's workers?

CHAPTER 23 TERRARIUM

I had brought a small terrarium with me to show the students.
'Terrariums are closed ecological systems and have been known to sustain for 60 years without opening up to the atmosphere. They demonstrate physical and living things surviving in mutual dependence. Once assembled with soil, water, plants and animals, all they need is sunlight. There are many designs on the internet and you can make one without much trouble or expense.

'A terrarium is a model of the World that shows interaction of the Earth with its atmosphere and living things. In a nutshell, photosynthesis takes place in green plants using sunlight and carbon dioxide. Cellulose and oxygen are produced which sustain respiration in animals, with energy, carbon dioxide and water. The two chemical equations are opposites and react mutually, benefitting each other. You need to learn these equations.

PHOTOSYNTHESIS takes place in the leaves, catalysed by chlorophyll.
sunlight + carbon dioxide + water → cellulose energy + water + oxygen

RESPIRATION takes place in the cytoplasm of every animal cell.
cellulose + oxygen → Energy + carbon dioxide + water

A catalyst is a substance that speeds up a reaction but is not used up.
Chlorophyll is a green substance in the leaves of plants.
Cellulose is plant sugar, like a celery stalk.

Cytoplasm in animal cells is a fluid containing organelles for respiration.'

'Why do we need to know this?' asked Buck

'It shows how livings things and the physical environment balance each other,' I said.

'Is respiration the reverse of photosynthesis?' asked Jessica.

'Yes; if you write the photosynthesis equation backwards, it is the same as respiration.'

'Not exactly,' she said. 'Photosynthesis has water on both sides.'

'Hmm,' I said. 'Maybe water isn't produced, but some has to be there for photosynthesis to occur.'

'Is life just a chemical reaction?' asked Jessica.

'It doesn't rule out other elements, such as happiness,' I said. 'It merely shows that because of this symmetry in the chemistry, the plants and animals can find a balance together. They provide for each other's basic needs.'

'I would like you each to make a terrarium,' I continued. 'You need a glass container that you can reach into, plant a moss garden and seal with a lid. The purpose is to watch it grow, demonstrating how this can take place in a closed atmosphere.

'Years ago, they used to put small animals such as wood lice or beetles into terrariums, to demonstrate they could live by inhaling the oxygen expired by the plants and from eating plant material grown from their CO_2 and excrement. The plants could get their CO_2 from the animals. Nowadays we don't imprison small creatures unnecessarily, but some might be in the soil and plant roots.

'Terrariums are decorative and an interesting exhibit.

'Next reader please.'

'It took hundreds of years of investigation and experimention to reach the understanding we have today of photosynthesis and respiration. The history illustrates the patience needed for scientists to investigate complex systems.

'Van Helmont in about 1600 grew a willow tree in a weighed amount of soil in a tub. At that time, it was believed that plants grew from substances taken from soil. He watered it daily, weighing the amount of water used. After five years, he discovered that the willow tree weighed about 74 kg more than at the start, As the weight of the soil had hardly changed, van Helmont concluded that plant growth could not be due to minerals, or anything in the soil and attributed it to his daily watering.

'Scientists' understanding shifted to believe that plants grew from water. The roles of carbon dioxide and animals were realized later.

'Jan Ingenhousz was a Dutch-born British physician and scientist who proposed the photosynthesis equation in 1772.

$6CO_2 + 6H_2O \rightarrow C_6H_{12}O_6 + 6O_2$

'It is the respiration reaction backwards, catalysed by chlorophyll.'

'What does the equation mean?' asked Tracey.

I explained the reactant atoms being transformed to product molecules.

'Sophie, please continue.'

'Lavoisier recognised and named oxygen in 1777.

'Photosynthesis was not demonstrated until 1796, when Jean Senebier, a Swiss botanist, pastor and naturalist, showed that plants absorb carbon dioxide and release oxygen with the help of sunlight. At night they respired.

'In the 1930s traditional beliefs in plant growth from soil were replaced with photosynthesis. In some hospitals, plants and flowers were removed from the wards at night, fearing patients might be asphyxiated, due to removal of oxygen by plant respiration.

'The slow progress from the first experiments to application of a theory illustrates the slow pace of investigation and understanding of climate science. The Earth's atmosphere is huge, too large for experiments. Theories have to be propounded with caution to avoid mistakes.'

'It took scientists 400 years to discover how plants grow. How long will it take them to be sure about climate change?'
'It could be hundreds of years before we are sure of causes and how to prevent them.
'Thank you reader.'

TASK 23
What part of the formation and operation of the Earth does a terrarium model?

TASK 23
A terrarium models solar radiation warming a closed system with atmosphere, soil, water and living things.

CHAPTER 24 SOLAR PANELS

I showed the class an A4 sized solar panel I had brought with me.
'This small solar panel has rechargeable batteries that have been charged in sunlight.'
I connected it to a reading lamp and switched it on.
'I bought it for $30 from a charity which distributes them to homes at subsidised cost in Africa. Many homes there are not connected to electricity. This panel enables children to do their homework after dark.'
'Too bad,' said Buck.
Everyone laughed.
'The solar panel was well received at a modest cost, a revolution in home life, where people were accustomed to sleeping after sunset.
'First reader, please.'

'Recent discovery of a technology for capturing solar energy and converting it to electricity has revolutionized domestic power supply. After subtracting 10% for reflection, due to albedo, or whiteness of a panel, typically only 15-30% of incident solar energy is converted to electricity. The remainder, up to 60%, warms the air in contact with the panel or is radiated away. Warming of the panel is reduced and less energy is reradiated when the solar panel is installed on a roof above a cooled interior.

'Solar panels have been said to reduce global warming, but if the 60% of energy from the Sun is reflected or radiated, wouldn't the reflected energy increase climate warming?' asked Kelly.

'I agree a panel could warm the atmosphere,' I said. 'Scientists have claimed that the heat that radiates from a panel is reradiated into space rather than lingering in the atmosphere.'

'How much of the re-radiation would warm the atmosphere?' asked Tracey.

'If there were clouds, perhaps a lot,' I said. 'Have you noticed how warm it is at night when there is cloud cover? Radiation bounces back from the clouds. Conversely, it can become very cold on cloudless nights.'

'Could environment warming be as great as from a coal-fired power station with 60% thermal losses?' asked Kelly.

'Possibly,' I said. 'Solar panels don't need the cooling towers and cooling ponds that thermal stations have. I suspect solar panels cause less convection and more re-radiation.'

'Next reader, please.'

'Coal technology is blamed for trapping heat. Although solar panels may heat the environment as much as generators, they do not produce carbon dioxide.

'Carbon dioxide is like other gases of the atmosphere and it is valued for its contribution to greenhouse warming of the whole Earth. The premise of renewable energy is that there is no combustion and less trapping of heat.'

'When it is cloudy, would the solar panels heat up and become less efficient? The turndown of output with increased cloud cover could reduce electricity supply severely. I heard a report that thick clouds cut electrical output dramatically.

'Solar panels are most useful at peak output when base-load generators can't meet demand. Near midday the panels can be complemented by electrical storages.

'Solar panels have other benefits. They are installed on roof tops and large tracts of land. The land is not entirely taken, for livestock can graze in the shade under the panels, which reduces evaporation, enhancing grazing.'

'The glare from a solar farm can be unpleasant. Hail storms and cyclonic winds can smash the glass. Solar radiation is most intense around midday and is nil after sunset. Clouds and rain reduce output. Whereas electricity is wanted at high voltages for transmission to consumers, panel output is at low voltage and transformers are needed. Consumption peaks at around 9 am and again at around 6 pm but there is not much panel output at those times. To spread out photo-electricity delivery so it can be consumed better, pumped storage and chemical batteries are used with solar panels.

Source: https://earth.org/solar-energy-facts/

'Installation of solar panels is subsidised or free in many places in Australia,' said Emily.

'Next reader, please.'

'The cost of a solar panel, compared with a coal fired installation, is about half for the same electricity consumption and the cost of coal has to be added on and the cost of batteries subtracted. Solar panels are attractive where they can be mounted on rooftops close to where the electricity would be used. They are most attractive at remote locations, avoiding the expense of transmitting electricity from a central power station. Solar panels need maintenance and eventually must be replaced.

'Although solar energy is said to be renewable, the Sun automatically replaces the radiation taken to heat the Earth. Two objects in space, like the Sun and Earth, keep a mutual balance in their temperatures, by emitting and absorbing radiation energy according to Kirchoff's Law. There is a solar balance between the Sun and Earth.'

'Is it the same as saying when radiation is absorbed as heat by a solar panel on Earth, additional energy will be taken from the Sun?'

'Yes, that seems right,' said Leblanc slowly.

'Then renewable energy can cause global warming,' I said. 'You won't hear salesmen of renewable energy devices saying that!'

'Would global warming of the Earth reduce or increase the Sun's solar energy warming the Earth?' said Buck.

'It depends on what has gone before. If the global warming is a priori, starting from nothing, then the Sun's contribution would increase. But if the Earth was already warmed up a posteriori, solar energy could be reduced.'

'So which is it?' asked Buck.

'Perhaps a bit of both. It's mutual.'

TASK 24

Make a list lists of advantages and disadvantages of solar panels for each of these locations, stating reasons.
- Houses in the remote outback.
- Commercial premises in outback towns.
- Houses in city suburbs.
- Apartment buildings
- Factories.
- EV charging stations.
- Coastal dwellings

TASK 24 Answers

Houses in the remote outback – low transmission distances.
Commercial premises in outback towns – larger rooftop areas for panel installation.
Houses in city suburbs – grid backup and sales possible.
Apartment buildings – group decision making needed.
Factories – panels unable to supply higher voltages needed.
EV charging stations – supervision of car owner competition in scarcity may be needed.
Coastal dwellings – weather could reduce supply.

CHAPTER 25 WIND TURBINES

'You will have listed advantages of solar panels. Today we are going to investigate wind turbines, which are sometimes considered as an alternative to solar panels.

'Next reader please.'

'Windmills have been used since the Middle Ages. Now they output electricity, with new designs and new materials. Construction is at a much larger scale and they are more available for communities to use.

'Installations in Australia are unlike those in Uruguay where almost every station, or estancia, has had a vaned rotor spinning on a pylon, charging batteries for mainly domestic use. There is usually a steady breeze. The installations are similar to wind pumps on farms in Australia, located near grazing, where livestock need watering. They have rotary vanes or windmills having high torques suitable for pumping water.

'Wind turbines in Australia are large, sufficient to power communities and small towns. Their large size enables them to reach up to stronger wind currents. They operate steadily by the inertia of a heavy rotor.'

'Is there wind everywhere?' asked Buck

'Australia is not known for windy conditions, but we have predictable wind at coastal locations and at certain times there can be cyclones,' I said.

'Next reader, please.'

Offshore wind farm
Source: Adobe Stock by Fokke Baarsen

Wind is air in motion. It is produced by the uneven heating of the earth's surface by the sun. Since the earth's surface has various land and water formations, it absorbs the sun's radiation unevenly with variable winds. The output of a wind turbine depends on the turbine's size and the wind's speed through the rotor. An average onshore wind turbine with a capacity of 2.5–3 MW can produce more than 6 million kWh in a year – enough to supply 1,500 average households with electricity.

'A wind turbine produces electricity 70-85% of the time, but it generates different outputs depending on the wind speed. Over the course of a year, it will typically generate about 24% of the theoretical maximum output (41% offshore). This is known as its capacity factor.

'Clearly, wind power is a highly variable. Energy storage is crucial,' said Leblanc.

'Next reader please.

'The role of the Sun in moving air was evident when I flew in a light plane at low altitude across a patchwork of agricultural land on the Darling Downs on a hot day in summer,' I said. 'It was a bumpy ride. When the land below us was pale coloured, with crops of ripened grain, our plane dropped vertically. Light colours are the best reflectors. Above dark newly cultivated land we ascended rapidly, because dark colours are the best absorbers and radiators. The dark surface would both absorb more solar radiation and emit more infrared, because good absorbers are good emitters. This heated the air above dark surfaces and caused it to convect a wind current that threw our plane upwards. The lower pressure below was drawing in cooler air horizontally and able to turn a turbine, but producing sickening vertical drops, above adjacent pale coloured areas.'

'Wind turbines can be damaged by high winds. They can overspeed, damaging the bearings, the blades or rotor can break and the tower or foundation can fail. In high winds, they feather the blades to prevent damage.

Australian wheat field. Source: Adobe Stock by denis_333

A pale land surface where radiation would be reflected and air would sink.

'Where would the heat go from above the pale land?' asked Norman.

'Less heat would be absorbed than by dark land. The air above would be less heated and the hot air carried along towards a current rising above dark land.'

'Would the energy flowing from cool to warm recirculate in convection cells?' asked Leblanc.

'Most assuredly,' I said. 'The hot rising air would expand, reducing pressure and then be cooled with altitude, moving along and sinking down.'

TASK 25
Which is a better electricity supply at a remote location, solar panels or a wind turbine?

TASK 25 Answer
A cost comparison of each with equal reliability, including batteries and equal storm resistances, could find variation in electricity demands and a difference between locations.

CHAPTER 26 AIR

'Last lesson, you visualized a bumpy plane ride caused by thermals.

'Solar panels and wind turbines operate in air close to the Earth's surface and are affected by weather. They do not have combustion products to pollute the air.

'In this chapter you will read and discuss how pollution can affect our air. We take air for granted but it mediates our climate and how we obtain our energy. Air is plentiful. We'll consider its quantity first, then its quality and finally how it is polluted.

'First reader, please.'

'Air wraps around Earth and the little that gets away into space is replaced from natural sources, keeping a pressure of 10.3 tons per square metre at sea level.

'The weight of the atmosphere is difficult to imagine. When you put your hand out of the window of a moving car, you get an inkling from the force on your hand that air is a weighty substance. We will watch a movie that demonstrates the heavy weight of air.'

I projected on their screens a steel oil drum standing on a barbecue grill, with Ken the scientific assistant pouring half a bucket of water into it.

He lit the barbecue to heat the bottom of the drum. After ten minutes I asked: 'What do you think is happening?'

'The drum is filling with steam,' said Tracey. 'It's escaping out of the bung hole.'

We could see a wisp of steam coming from the top.

'Correct. The bung is not yet screwed in. Anything else?'

'Air is being flushed out by the steam.'

'Yes. You can't see the air, but you can see the steam condensing in the cool air outside. Air in the drum has expanded and been pushed aside by steam.'

Ken turned off the barbecue and wearing gloves quickly screwed in the bung.

He sloshed a bucket of cold water over the top and sides of the drum.

Then he stepped back.

'Do you notice anything?' I asked. 'I can hear a clicking and sucking noise and the drum has moved a little, as if the metal is being bent. What's happening?'

'The steam in the drum is condensing,' said Tracey.

'Correct. What will fill the drum when the steam has all condensed?'

'Nothing,' said Leblanc. 'A vacuum.'

Suddenly, there was a bang and the drum collapsed into a crumpled mass. For a moment the students' reactions were fearful. It teetered, then fell off the barbecue with a clang.

'What happened?' I asked.

'Atmospheric pressure overcame the vacuum in the drum and crushed it. You have seen what air at 10.3 kilograms per square metre can do when it is unopposed.'

'Cool,' said Buck.

'That's all, I said.
'Continue reading.'

'Aristotle declared that everything consisted of the elements: earth, air, fire, and water. He thought the heavens were made of aether, were weightless and incorruptible.

'In 1667 combustion was thought to release an invisible substance, phlogiston. But this was a false trail. Oxygen was discovered and named by Lavoisier in 1777.

'The concentration of carbon dioxide pollutant is nothing like a 'veneer' or thin layer that Gore likened it to, as if its higher density would keep it down. Perfect mixing of gases is a physical law,' said Tracey.' Approximately 99% of Earth's atmosphere is contained within the region extending up to about 50 kilometres above the Earth's surface. This includes the troposphere, which extends in a thick layer up to about 12 kilometres and contains about 75% of the atmosphere's mass. The stratosphere extends from about 12 kilometres to around 50 kilometres and contains most of the remaining 1%. Carbon dioxide is not like 'an eggshell', holding in radiation reflected from below. It mixes itself throughout the atmosphere. It is a very small percentage, 0.042% by volume, of the troposphere and stratosphere, which may be heated by atmospheric gases circulating by convection.'

'Is there really so little carbon dioxide?' asked Leblanc.
'In past ages, at times there has been much more, but there is no evidence of disastrous consequences,' I said.
'People are misinformed or dishonest that carbon dioxide is now at crisis level.'
'Yes, I think so too.'
'Next reader, please.'

'Air plays a vital role in the lives of living things. In a closed room, a person will take several days to use up all the oxygen. The waste carbon dioxide accumulates. Trapped miners and submariners will begin to feel unwell and weak. When there is insufficient air, it is like visiting the Himalayas at high altitude, or the Andes, or Kenya, where air pressure is lower at high altitude. Sherpas in Nepal are habituated to thin air and athletes train at high altitude camps in Kenya to increase their endurance when they race at sea level.
'Winds are supposed to dilute 'bad air' from swamps, such as methane, giving the city of Buenos Aires its name, meaning good air. Mosquitoes are disease vectors that exploit air for spreading malaria, meaning bad air.

'Air quality varies from place to place,' I said. 'I recall arriving in Brisbane from the UK and being amazed by the balmy summer weather. I had never experienced such humid warm air before.

'In Edmonton, Canada, in winter, I had never experienced such dry coldness, nor had my sinuses bled before.'

'Read on, please.'

'Most weather in Australia falls within predictable limits, due to large slowly moving air masses. Irregularly there can be discomforting events such as floods, droughts, cyclones and bushfires. People cannot afford to completely protect themselves against disasters and risk losing everything. Compensation by insurance and public authorities covers some major losses,' I said. 'For home buyers, the weather is never predictable enough. A disaster is always possible.'

'Pollution of air has gained public attention. The Earth's air has acted as a free dumping ground for human wastes. Asbestos and artificial stone dust have been prohibited. Air has a self-repairing and self-maintaining action that has been sufficient for the atmosphere to remain in public ownership, without sovereign nations claiming it. But the access to the body of pollutants, through airways and the respiratory system, is being protected in developed countries.

'The atmosphere is complex and understanding is confounded partly by fake science and false theories, in my opinion, as you learned when we studied the Greenhouse Effect. In our next chapter, you will find out about how various pollutants have been discovered.'

TASK 26
What is supposed to be the problem with 0.042% carbon dioxide in air?

TASK 26 Answers
It has been associated with climate change events and global warming.

CHAPTER 27 POLLUTANT EVIDENCE

'Complex systems are difficult to understand and slow to analyse. Understanding takes time. Causes of climate change cannot be observed and must be inferred or deduced. Actions based on partial analysis may do more harm than good.'

'In this chapter I want you to compare a pollutant attributed to climate change, carbon dioxide, with some well-known problems of pollutants.

'First reader, please.'

Smog used to cause deaths in cities and led to the banning of coal burning in homes and industries. The word 'smog' was coined in the early 20th century, and is a portmanteau of the words smoke and fog to refer to smoky fog due to its opacity, and odour. Smog was reduced by control of coal burning in cities but has continued in densely populated locations such as China, India and Indonesia. The improvement encouraged other air quality controls.

'Mercury poisoning at Minamata Bay, Japan, in the 1950s drew attention to the possibilities of chemical pollution of water. The source was a factory dumping waste at the coast, only slowly realized from accumulating birth deformities.

'In 1958, Rachel Carson received a letter from a friend in Massachusetts bemoaning the large bird kills that had occurred on Cape Cod as the result of spraying with DDT to kill mosquitoes. 'She campaigned to ban the use of DDT in the USA and investigated conservation of ecology in her book 'Silent Spring'. DDT was a chloro-carbon she said magnified weakening of predator birds' egg-shells in the avian food chain.

'Carson became famous for her view that indiscriminate application of agricultural chemicals, pesticides, and other modern chemicals polluted our streams, damaged bird and animal populations, and caused severe medical problems for humans. DDT ceased to be used for anti-malaria spraying world-wide.'

'What was the evidence of harm to humans?' I asked.

'Carson's friend said there were kills of large birds, such as eagles,' replied Kelly. 'But no humans.'

'Next reader, please.'

'In the many years since, although the biomagnification she described made her famous, there were critics who claimed that she had never observed weakened egg-shells in birds of prey. Some of Carson's other findings were verified, however.

'If DDT spraying for controlling malaria in African countries and elsewhere had continued, Carson's critics claimed many people who died from malaria might have lived. No effective substitute for DDT was found, even after many years. The use of DDT is reported to have resumed now in some countries, without report of further large bird kills.'

'This information suggests that Carson acted irresponsibly, without sufficient evidence of DDT biomagnification in birds,' said Leblanc.

'The banning of DDT from pest control before substitutes were available was unfortunate,' I said, 'especially in places with malaria.'

'Read on, please.'

'A different researcher investigated tetraethyl lead, added to petrol to improve its anti-knock quality. After many years, lead poisoning of members of the public was detected. Use of lead in petrol was banned. Confidence in finding and eliminating harmful pollutants was growing.

'A gas tragedy at a chemical plant at Bhopal in India in 1984 was an early experience of fatal air pollution. The final death toll was estimated to be between 15,000 and 20,000 and some half million survivors suffered respiratory problems, eye irritation or blindness, and other maladies resulting from exposure to the toxic gas.

'Accidental water pollution by oil spillages from Exxon Valdez in 1989 and from Deepwater Horizon in 2010, were intense locally and killed wild animals.

'A more puzzling case was the Ozone Hole over Australia. A 'hole' in Earth's atmosphere was found in the 1980s,' read Tracey. 'Scientists thought chlorofluorocarbons (CFCs), long-lived chemicals used in refrigerators and aerosol sprays since the 1930s, could be catalysing the disintegration of ozone in the hole. The theory was that air pollution by refrigerant gases thinned the ozone layer and allowed in shortwave radiation causing skin cancer.

'In 1987, an international treaty called the Montreal Protocol, phased out the production of CFCs to mend the Ozone Hole over the Antarctic. Scientists observed a 20 percent decrease in ozone depletion during the winter months from 2005 to 2016, the first definitive proof of ozone recovery.'

'There seems to be some doubt that the CFC's fixed the ozone problem,' said Tracey.

'When the hole filled in, it was assumed that stopping refrigerant gas emissions had done it, but it could have happened anyway. There is some evidence people have suffered fewer skin cancers. But the cause of the Ozone Hole remains uncertain, because it forms and disappears on an annual basis, in springtime over Antarctica.'

'Jesus, is anything a sure thing in science?'

I shook my head.

'We don't hear about the easy investigations,' I said.

'Next reader please.'

'When Chernobyl power station suffered an explosion in 1986, the radiation and radioactive fallout were perceived by scientists as warranting evacuation of the area and they declared an exclusion zone at a radius of 30 kilometres. There is consensus that a total of approximately 30 people died from immediate blast trauma and acute radiation syndrome in the seconds to months after the disaster, respectively, with 60 in total in the decades since, inclusive of later radiation induced cancer.

It was expected that wildlife remaining would be badly affected, but by 2020 it was evident that the wildlife had not suffered adverse effects and had re-established conditions prevalent before the disaster. Effects on the evacuated human population were much less than was expected.

Tobacco was suspected as a cause of lung cancer and a long confrontation with cigarette manufacturers eventually resulted in smoking, tobacco and vaping restrictions.

Asbestos fibres were linked to lung cancer and eventually the substance was banned for use in buildings and some buildings such as schools had to be rebuilt.

Alcohol has long been suspected of injuring health and causing addiction. It's use is controlled but it continues to be widely used. It does not have the status of a pollutant yet.

The experience was that substances polluting air had uses contested by their manufacturers until the scientific evidence was unequivocal.

'It became clear that the effects had been guessed even by experts and had been wildly pessimistic,' I said. 'Can we learn anything relevant to climate change?'

'If the pollution at Chernobyl was so exaggerated, we should be cautious how much pollution effect we attribute to carbon dioxide,' said Leblanc.

'It will be difficult to demonstrate carbon dioxide is a cause of climate change. Circumstantial evidence has not been enough to condemn other suspected pollutants.'

Next reader please.

'Experimental validation of a climate effect by reducing carbon dioxide emissions in air has not been available but alarmism and anti-carbon hysteria have been taken up by school children and others, who do not understand the important role of the carbon cycle in plant growth or of carbon as a relatively benign constituent of organic industrial chemicals used in everyday products. Carbon dioxide's so-called pollution of the atmosphere is a misnomer, for its effect is supposedly thermal, not toxic, nor chemical and the evidence is possibly circumstantial.

'Read on, next person.'

'48 years after Silent Spring, Al Gore in his movie An Inconvenient Truth, presented an emotional attack on carbon dioxide as a pollutant causing global warming. Today, 18 years later, Al Gore's doomsday predictions have not eventuated. Dr Tim Ball (d.) described Dr Michael Mann as a liar for exaggerating the global average temperature increase in the movie and fought a defamation action that was dismissed unresolved because the court case had dragged on for many years and both the litigant and the defendant were aged and having difficulty attending the court in Oregon.

'There is ample evidence that climate change has been exaggerated and Gore acted irresponsibly,' said Jessica.
Next reader please.

'Gore besieged conventional energy authorities with his claims and fostered images of immanent climate disaster, triggering a changeover of electricity generation technologies that has made technology suppliers huge profits and their directors large personal fortunes.'

'Gore was famous and powerful. He was in a position that could bully administrators and scientists into rejecting fossil fuels, at great cost to their communities. It was carried by belief and politics, not science.'

'It could seem possible to the uninitiated that whether carbon dioxide causes global warming or not can be decided now. In practice, there is no simple test because the scale is large and scale models do not have the complexities of the atmosphere. It is impossible to establish 'truth' when causality, observation and measurement cannot be achieved. Empirical proof is replaced by weight of opinions in the postmodern way and refuted by others who claim it is a hoax.'

'Do you believe it was a hoax?' asked Jessica.
'No. It seems to me that better understanding of carbon dioxide is needed before such an important issue can be decided,' I said. 'But it is an unlikely pollutant.'

Next read on.'
'Climate events, heatwaves, bushfires, droughts, floods and sea level rise have been attributed to air pollution without conclusive evidence and with much refutation. Carbon dioxide has been claimed to be the main culprit without validation. Methane in tiny quantities was also blamed on cows, thawing of permafrost and livestock. Breeders are searching for fartless breeds and cattle foods such as seaweed that could reduce flatulence. The amount of carbon dioxide in air is tiny and cannot possibly account for large climate effects claimed. Carbon dioxide is the main food of plants and the current concentration is historically low. Moisture is much more significant for trapping heat, but is omitted from most calculations.'

'One of the pleasures of life is a fart.' said Kelly. 'I don't like it that they are making cows miserable.'
They laughed.
'Cows have more couth than you, Kelly.'

'Remembering the fiascos with other pollutants, my hope for the climate change problem is for honesty, reason and caution. Science is always sceptical and agreement is not achieved by consensus nor by edict of majorities. The frequent announcements of disasters have more to do with controlling fearful masses, who are unable to analyse the welter of misinformation from governments, media and corporations, expecting them to supply relief in return for votes and taxes. The weight of public opinion seems to be ultra-cautious and in need of reasoned rationalisation to offset tendencies towards totalitarianism.

TASK 27
What are some air pollutants that have had large scale treatment strategies?

TASK 27 Answer
    Smog
    Lead in petrol
    Sulphur dioxide
    Carbon dioxide
    Tobacco
    Asbestos

CHAPTER 28 ECOLOGICAL RESPONSE TO CLIMATE CHANGE

In this chapter you will consider whether humans and other species can adapt to likely new climate conditions. Long term thinking is needed. It opposes those who want to resist climate change at any cost. Perhaps humans can compromise and adapt their homes to counter hot or cold conditions while they last.
'First reader, please.'

At the end of the last ice age, when the ice retreated towards the poles, woolly mammoths might have preferred to adapt to coldness and the lifestyle in which their hairy coats had evolved to be most comfortable. Is that plausible?

'Yes,' said Tracey. 'Polar bears today are having difficulty finding their traditional foods of seals and penguins as the ice melts, Will they die out? Is it possible that they will have to adapt to a warmer clime and different foods?'
'In warmer or cooler climates, farm pests could flourish, in crops whose abundance would be different,' said Biggs. 'Should we protect crops from climate change that would expose them to pests and competition from other species.'
'There isn't a problem!' said Norman. 'There has been climate change since forever and we still have enough species. It is natural for some species not to survive. Darwin expected there would be extinctions. People who demand everything to stay the same make me puke. It's so naïve. Of course, a few species will have difficulty adapting, but they leave openings for others to fill in and new species

to evolve. The only relevant observation is that the World is continuing, changing slowly. Those who say otherwise are deluded.'

'You may not care about losing species, Norman, but plenty of people do!' said Leblanc. 'Humans are having such large effects, we can't sit back and hope it will turn out okay.'

'It was on the news that migratory birds, terns, have not arrived this year,' said Tracey. 'They said it was due to climate change. Should we turn a blind eye?'

'Maybe the birds have changed to a different route,' said Jessica. 'When conditions change, animals adapt.'

'One good tern deserves another,' said Kelly.

'Hold on,' I said. 'Aren't you getting ahead of yourselves. Don't you need to find out how much climate change could be a problem and how soon that could occur. Any action should be preceded by observation, measurement and analysis. You can't simply jump into the deep end before you know how deep the water is or how to swim.'

'Farmers in Iceland and Newfoundland like warming and retreat of the ice. They can cultivate new areas,' said Leblanc.

'Warming is not bad for all species in all places,' said Biggs. 'There can be advantages. They say that photosynthesis speeds up at higher temperatures and we can get higher yielding crops and greening of deserts. Would that not be good?'

'Our people can't act for all the planet's people and species,' said Tracey. 'Humans who live in hot climates could want to oppose the discomfort of warming. In some cold countries they could want to maintain ice and snow, for recreation and tourism.'

'It doesn't seem likely that we could arrange for every species to get the climate change they want,' I said. 'That could be why environmentalists oppose all change.'

'Humans have put up with weather abnormalities, causing droughts, bushfires and extreme events, since forever,' said Norman. 'Why can't they now put up with occasional climate change?'

'I like it that you are suggesting humans can adapt,' I said. 'Are we really flexible enough to adapt to climate change?'

'Some animals can move, seeking out microhabitats that remain in their preferred temperature range,' Norman said. 'Or they can adjust their behavior to be more active at cooler times of the day, which might buffer them against the effects of climate change. But behavioral changes can only be pushed so far.'
'Next reader, please.'

'Owls have wings that allow them to fly around their habitat, with eyes adapted to help them see better at night, with brown and white patterns on their feathers that help camouflage them against the background. Barred owls use these features to live in a forest habitat full of trees. But when the forests are cut down, they may not be able to adapt anywhere else.'

'It's difficult to predict how much climate change is coming and what the effects on the species will be,' said Leblanc. 'Your owls, Norman, seem to be accustomed to an environment that could quickly change. I'm not sure if coral has such needs. Probably because I am not familiar with coral habitats, they seem to me to have sufficient slack to accommodate climate change.'
'Maybe corals aren't really slack,' said Buck. 'They could be faking it.'
'They might not be able to adapt,' said Leblanc, 'because they are too highly strung.'
There was laughter.
'Their mutualism could be strained.'
More laughter.
'They need counselling.'
'Can humans adapt?' I asked.
'They can develop crop varieties that are more tolerant of heat, drought, or flooding from heavy rains,' said Tracey, 'to provide more shade and air flow in barns to protect livestock from higher summer temperatures.
'Many people think climate change has to be opposed,' said Kelly. 'They want governments to act. What makes me vomit are changes

applicable everywhere, as if they would be subscribed to by people everywhere and would benefit all the species. There has to be planning that benefits the majority of people in various locations.'

'Humans have ridden roughshod over environments,' I said. 'Climate change is gradual, accumulating results insidiously. Many people want more response, not ignoring it. They want an elegant sophisticated solution that will permit an economic recoil. Something like Net Zero.'

'Like stopping change altogether?' said Biggs.
'Fat chance,' said Norman. 'Humans have survived change in the past and will again.'

TASK 28
Answer this question.
Can humans distinguish climate change from evolutionary change?

TASK 28 Answer
With difficulty.

CHAPTER 29 LIVING IN SPACE

'Humans seem willing to consider emigrating through space and settling on another planet. In this chapter you will work in teams investigating some possible living conditions in space and write a plan for the journey and settling in on arrival. The huge distances and long journey times require new technologies and special training.
   'First reader, please.'

   'Your task is to inquire into Living in Space. These notes will guide your work in small groups to submit a plan for publication specifying conditions on a 200 passenger spaceship, to go to a planet in constellation Centaurus, orbiting around its sun. Alpha Centauri is a triple star system located 4.3 light years from Earth. It's the nearest star system to our sun, at 4.159×10^{13} kilometres. The Space Launch System would be ready in 2028 and is expected to reach 20% of light speed of 300,000 kilometres per second, or 216 million kilometres per hour.
   The journey time to reach Alpha Centauri is calculated as follows.
Time = Distance / speed
Distance = 4.159×10^{13} kilometres
20% of light speed = 216 million kilometres per hour
Time = $4.159 \times 10^{13} / (2.16 \times 2.4 \times 3.650 \times 10^{11})$
= $4.159 \times 10^{2} / 18.92$
= 22 years

   'Would it be possible to take everything I would need for a 22 year journey?' asked Jessica.
   'Your team will decide what you would need and what you could do without,' I replied.

'Next reader, please.'

The long journey could create special problems for people of different ages and you will need to identify how everyone can be kept healthy, active and positive. Anti-group behaviour may need to be supervised and prevented from causing harm.

Below is an artist's conception of a starship – called the Starshot Lightsail – accelerating away from Earth. The craft uses an Earth-based laser array to get up enough speed to travel to this next-nearest star. The shape and composition of the new solar sail were recently redesigned, so the sail won't melt or tear during the acceleration phase.

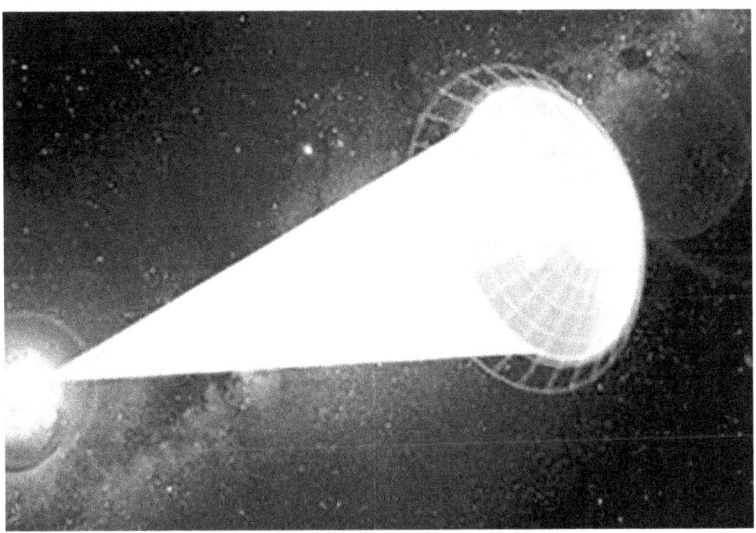

Source: Masumi Shibata Breakthrough Initiatives, CNET

TASK 29 OUTLINE
Conditions there will be harsh: the atmosphere is thin carbon dioxide, cold and irradiated with UV from the Sun, but there is water at the poles and food could be grown. You will plan everything you will need to take and do on the spaceship, for everyone to live reasonably well.

The plan is to be 1000-1500 words in length and include appropriate lists, tables, graphs and diagrams.

INQUIRY TOPICS
Your plan should consider some of the following.
1. Travel conditions
   Passengers will endure 22 years of continuous travel in a spaceship which needs to be comfortable but low weight, including fuel for the journey and possibly for return. Could people withstand the conditions? Is there anything they can do to prepare to adapt?
2. Oxygen
   Breathable oxygen can be manufactured by the electrolysis of water.
3. Nutrition
   The spaceship must carry enough food for the journey with little space for growing anything. Suspended animation could be available but may be risky. What diet would people need to remain healthy?
4. Medical care.
   Passengers can be of all ages. The scope could include treating accidents, preventing disease, birthing and birth control. Passengers would ideally be fit and have skills that would contribute to the group during the journey and on arrival. Can epidemic disease be avoided?
5. Waste disposal
   Exhaled air, body wastes and disposal of any dead bodies must be done safely. Water can be recovered from waste and recycled. Could waste disposal be prevented from contaminating space?
6. Energy supply
   The spaceship systems would be powered by electricity from solar panel arrays. As distance from the Sun increases, there will be less solar radiation but as you near the destination, the panels could be turned to collect energy

from Alpha Centauri. How can energy and electricity demand be minimised?
7. Water supply
All water must be carried from Earth and recycled in case the destination does not have any accessible. How could washing and showering be minimised?
8. Gardening
On arrival, crops could be planted to grow food. What crops could be grown in Mars-like conditions or would terraforming have to modify it to a more habitable atmosphere, temperature and ecology?
9. Accommodation
The travellers could take large tents to contain living spaces on the surface, in an artificial atmosphere, or wear space suits with helmets.
10. Heating and cooling

Chapter 29 has information about refrigeration and heating systems.

Meet with others in your class, forming groups with a leader and following common interests to share planning for 200 passengers of all ages.

ENTER YOUR NAME FOR TWO INQUIRY TOPICS, AT LEAST ONE AS LEADER

INQUIRY TOPIC	LEADER NAME	FOLLOWER NAMES
Travel		
Nutrition		
Medical care		
Waste disposal		
Energy supply		
Water supply		
Gardening		
Accommodation		

HOW TO CHOOSE A TOPIC FOR PLANNING
1. Select a topic you would like to assist in planning.
2. Contact others with the same topic interest.
3. Consider the scope of planning.
4. Plan elements in which you have relevant experience.
5. Select a topic leader. You could put yourself forward.
6. Contact other teams and find out any plans that impinge.
7. Finalise your topic plans, allowing for other teams' plans.
8. Contribute to writing your team's planning document.
9. Submit it.

'How can we plan for conditions never experienced before?' asked Mark

'You will have to predict the worst conditions to design for, stating your assumptions,' I said. 'For example, what is the least and greatest gravity you could encounter and what equipment will you need?

TASK 29 ASSESSMENT CRITERIA
Your team will submit an article for publication with a plan for human living at the destination. The article is to be 1000-1500 words in length and include appropriate lists, tables, graphs and diagram.

You can work on this project in class. Request a breakout room for your discussions. At the end of the second lesson, submit your topic nomination form, to be shared with the other class members. The plan is to be submitted after three weeks.

CHAPTER 30 REFRIGERATION AND AIR CONDITIONING

Air conditioning is a basic technology in space travel and must be thoroughly understood by space travellers who use it and maintain it properly.

Your inquiry in Chapter 28 will need planning of spaceship cooling and heating using an air conditioner. Planning and communicating ideas for a mechanical system requires scientific understanding and strong communication skills.

This is a self-study topic, for you to read and apply to Task 28, finding answers to the questions from the internet.

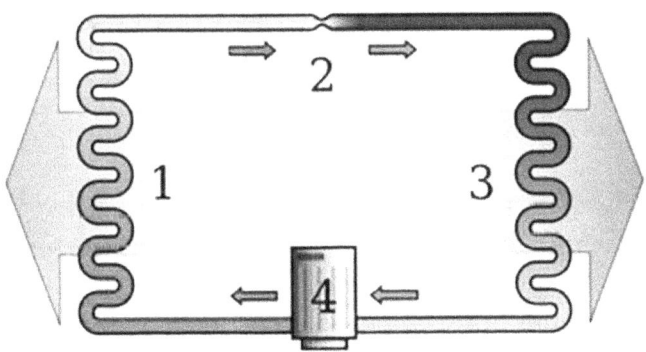

Source: Wikipedia

1 – Condensing coil 2 - Expansion valve 3 – Evaporator 4 – Compressor

When a substance changes from the liquid state to a vapour it is said to have evaporated.

This occurs above a certain temperature and takes energy known as latent heat.

When the liquid is under pressure, evaporation is at a higher temperature, requiring more heat.

Outside a refrigerator, the refrigerant gas is compressed and warms up, radiating heat from the condensing coil, changing to a liquid. The liquid flows through a tube inside the refrigerator, through the expansion valve, where it evaporates to a gas, gaining heat from the refrigerator contents, which cool down.

'Why is it necessary to heat the refrigerant?' asked Norman. 'Wouldn't it make more sense to cool inside directly?

'Agreed,' I said. 'But there might not be anything colder available? The refrigerant cools by compressing the gas, cooling it outside, then expanding it, lowering the temperature inside.'

TASK 30
Find answers to these questions.

1. Draw a diagram showing the parts of a refrigerator and colouring the condensing coil red and the evaporation coil blue.
2. Show the position of the compressor and the expansion valve.
3. How do a refrigerator and air conditioner work?
4. To where does the heat go inside the refrigerator?
5. To where does the heat in the condenser go?
6. Where does the heat go from the evaporator?
7. What causes the refrigerant to circulate through the condenser and evaporator?
8. Free cooling can be selected when air outside is cooler than the internal air and the compressor does not need to be used,
9. Explain why an air conditioner can heat a space in winter.
10. Explain why a refrigerator must have an air space behind it?

11. When chlorofluorocarbon was used as a refrigerant, what problem was there?
12. What refrigerant is used today?
13. Is the ozone hole present today?

Check your answers with others in your team.

CHAPTER 31 BRISBANE CIRCULAR UNDERGROUND RAILWAY

'Transporting of humans and settling them in colonies in space, could be planned by authorities experienced with making similar provision on Earth.

'Governments are responsible for transporting their citizens. Brisbane City Council has a public transport system serving a population of 2.5 millions. Compared with energy and water systems, a public transport system is more extensive and transparent. Expectations are demanding.

'Read this chapter taking turns, studying the information for in-class discussion.

'First reader, please.'

'If you live in or visit Brisbane you have probably travelled around by car like most people. You may have used public transport: buses, trains and ferries. Active transport is also available, using bicycles, scooters and walking. Were you able to use public transport as much as you wanted?'

'Why does city traffic have mostly single-passenger vehicles and scarcity of parking,' asked Tracey.

'Is it because car transport is more exclusive and preferred by snobs?' asked Buck. 'They wouldn't be seen dead on public transport.'

'Buck, don't use emotional language.'

'Those people deserve to be ridiculed.'

'They are entitled to travel as they want provided they are not breaking any laws,' I said.

'Another problem with cars is that their owners want to park them close to their workplace with everyone else, so that however many parking spaces are built, it is never enough,' said Kelly.

'Next reader, please.'

The challenge in this chapter is for Brisbane City's government to create an integrated transport strategy to coordinate six modes of transport (walk; cycle; ferry; bus; rail; car) managed by 3 levels of government: Commonwealth; State; City Council; in a wider area approximately 60 kilometres in diameter; with a population expected to live, work, study and recreate, increasing to about 5 millions by 2050 and 7 millions by 2070.

'The government has to provide transport alternatives, to all residents fairly.

'The Queensland Government's policy standard is that 90% of dwellings should be within 400 metres of an existing or planned public transport stop. In 2018 only 61 percent had such a stop and only 12% had services running every half hour. The Brisbane City Council operates efficient bus services, but they aren't available close enough to where most people live. The other thing is the roads are jam packed with cars most of the day. Buses with 30 people aboard have to wait for lines of single passenger cars to move out of the way. Most of the Council's money is spent improving the roads, but once a bottleneck is widened, another jam results further along.

'Next reader please.'

'Transport Hubs are where passengers board or leave transport for origins and destinations throughout wider catchment areas. At Southbank, a de facto hub has grown like topsy, more as a congested river-crossing bottleneck, than as a premeditated hub providing location sharing, user convenience, and technological efficiency.

'A new Brisbane hub is required to move the growing number of passengers across the Brisbane river, removing the burden of connectivity from the Boggo Road to Roma Street transport corridor and from the Southbank bottleneck nexus. An underground railway circle line, like London's Circle Line, is proposed to replace it, functioning as a router, forwarding travellers between different parts of their journeys.'

'To avoid convergence of transport at the Southbank hub, could there be two hubs, one on each side of the river?' asked Norman.
'That would double the congestion,' I said. 'Nice try, Norman.'
'Next reader, please.'

'Keeping the hub in the centre could exacerbate the problem. A circular rail route, like the one below, is the shortest way to connect a population, providing close access and enabling trains to run in opposite directions, doubling service frequency. Without corners, trains can run fast, with minimum maintenance from friction at bends. As circle diameter is increased, construction cost increases with circumference, but the population served increases at a similar rate, even allowing for thinning of residential density. Thus, benefits per resident from an underground railway do not diminish with increasing radius. The constraint on size of the circle is availability of capital. Extra dollars spent on a wider circle would not reduce the rate of return on investment.
'The map below shows how movement through the city is opposed by the river.
'The meanders of the river cause traffic congestion.

'The biggest problem with roads is the river,' read Kelly. 'The city is slap-bang in the way of the river and there aren't many bridge crossings. 'If river crossing was simply a matter of one bridge, it would be quick, but the Brisbane river meanders around, taking its time, holding back floodwater and submerging roads. You can't go anywhere without getting tangled up with roads radiating out from bridges. Convergence near the city centre makes commuting crowded and unpleasant.
'Would a Circle Line improve travel in Brisbane?'

'Yes, many people would benefit. Although the circle would be near where more people live, some would need to take a bus or car to a station.'

Next reader, please.

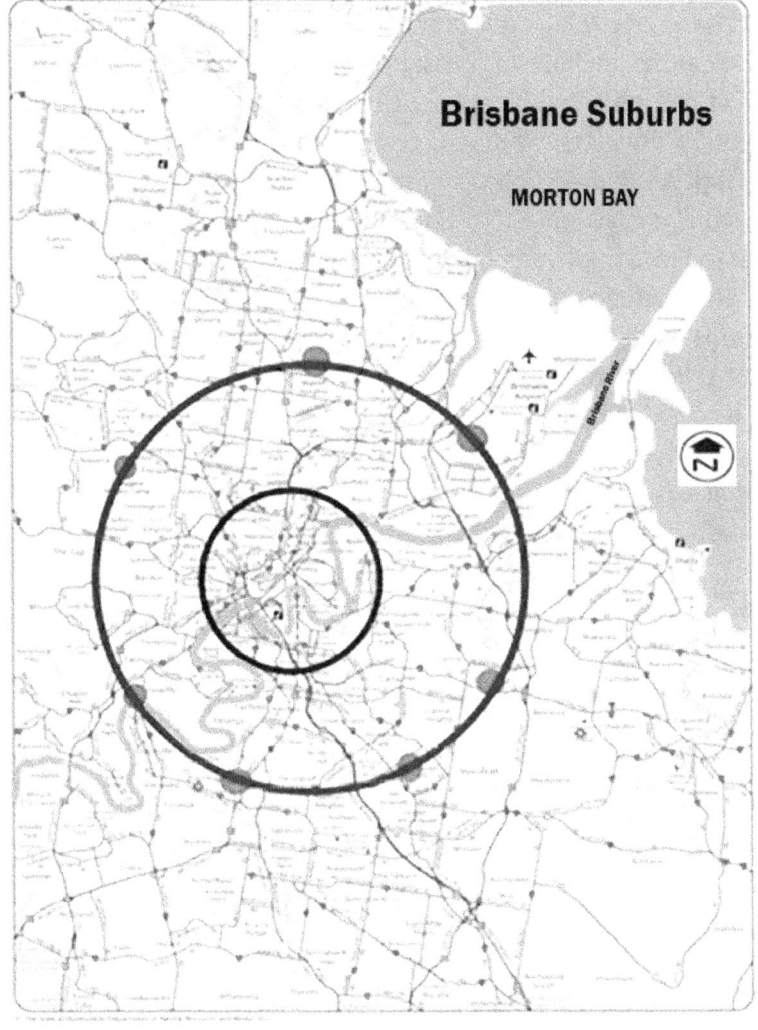

Brisbane Circular Underground Railway (courtesy of The State of Queensland.)

'On the previous page is a map with my proposal for a circular underground railway (black). 'The Circle Line is proposed to be the distributing hub of an integrated system. Brisbane needs to grasp the nettle with a staged multibillion dollar underground railway development.

'The inner circle is an alternative railway proposal to go around the city centre, included for discussion purposes.

'The red dots are transport-oriented development centres, on a circle of 8 kilometres radius crossing the river through two tunnels. The stations would be at Indooroopilly; Enoggera; Chermside; Yeerongpilly; Garden City; Carindale; and airport. They would be connected by busways to the present City Business District. Travellers could get from anywhere to anywhere, going in one of two possible directions. When the tube reaches their destination suburb, they can walk or take a local bus. There will be rapid transit to the City centre along a corridor from each hub.

'Will it be built?' asked Leblanc.

'Yes, if the city government plans for its citizens' future. It connects many people together who have been without a public transport service. Polycentric Brisbane is already underway, with nomination of TOD centres.

'It creates new living places, new businesses and new jobs. But it would be expensive and take years to build.'

'What would be the cost?' asked Kelly.

'My guess is 30 billions for the first phase, which is the circular line without stations.'

'Does it have to be underground?' asked Biggs.

'The city is already built up near the city centre, and it would be cheaper to tunnel underneath. In other places, it may be possible to construct an overground track, or one in cuttings, partly underground.'

'Why does there have to be a circle?' asked Jessica.

'It would act as a hub, gathering travellers from outer suburbs, enabling them to travel around to anywhere in the city in one journey. The diameter is 16 kilometres, around the most densely populated city and avoiding congestion at the current hub, Southbank, where travellers are forced to converge.

'I like it,' she said.

TASK 31

'Describe the circular underground railway proposal, stating advantages and disadvantages for travellers. Describe energy saving possible, with the black outer circle at 8 kilometres radius. Could the system bring public transport to everyone, within 400 metres of an existing or planned public transport stop, at half-hour intervals?'

CHAPTER 32 ENERGY TECHNOLOGY

'I begin this chapter by drawing your attention to revolutionary change in energy supply and consumption. Australia's power stations used to be her largest industry, larger than steelworks, aluminium smelters and oil refineries. Australian coal's heyday, as the fuel of power stations, railways, steamships and factories, has disappeared. Australia has prospered from dumping minerals into ships for export, mainly coal and iron ore.

'The purpose in this chapter is to review current development of energy technology in Australia.

'First reader, please.'

'Queensland's abundant coal gives the State a monopoly status like Saudi Arabia's for oil production,' a mining colleague told me. 'There is sufficient coal for domestic and export demand for hundreds of years.

'But coal export is under threat from environmentalists and only a few old coal-fired power stations operate, as islands of big coal technology. Coal mining and steam raising are vilified as old-fashioned, dirty, dusty, dangerous and unhealthy. The industrial revolution that familiarised Europeans with coal technology was less familiar to Australians, who were suspicious of heavy industry and regarded power stations as workplaces where employees were exploited, endangered and shamed for emitting pollutants.'

'Is coal technology olde worlde and stigmatised?' asked Tracey.
'Yes,' I said.
'Next reader please.'

'The electrical output end of power supply is modern and clean, enabling customers to enjoy electrical lifestyles, with electricity reticulated silently and devices operable at the flick of a switch. Climate change has contrasted clean electricity with stock images of coal fired power stations belching pollutants from cooling towers. This is false, because cooling towers emit only harmless water vapour.

Coal railing and handling are accused of being 'dirty' despite continuous dust suppression. Machinery operators at mines enjoy radio contact from air conditioned cabs. The jobs are well paid and safe compared with work in less technological environments.

The Net Zero intervention aims to cut carbon emissions, by 2050, with additions to the atmosphere to be less than removals.

'What does Net Zero mean will happen?' asked Kelly.
'New fossil fuels technology will only be allowed when old technology is shut down,' I said.
'Will this prevent growth of fossil fuel use?' asked Kelly.
'Correct.'
'What about export?'
'I'm not sure.'
'Read on, please.'

'Net Zero critics of coal-fired power stations often do not understand the technology but this does not daunt them from their hatred of carbon dioxide, a benign by-product essential for plant growth. Such is the insidious misinterpretation of carbon dioxide, it is feared by novices who without chemical understanding regard it as toxic. There is much in electricity supply that is unfamiliar to beginners.

'Coal is the subject of a vigorous debate between those who value it as an important economic commodity, opposed by those who claim it is causing climate change.'

'Coal has been relegated from Australia's largest export commodity and major revenue earner, to 'has been status', on the say of Al Gore's self-promoting lies, with only distant prospects of any benefits.

'I don't think you should accuse Gore of lying unless you have evidence,' said Pamela.
'His theory of carbon dioxide pollution is wrong,' I said.
'Has it been tested?' she asked.
'It's not possible to conduct a global warming experiment, because the scale is too large.'
'Then you cannot say it is untrue?' Pamela objected.
'I can say it is untrue because although it lacks empirical evidence, it is inconsistent with other theories of atmospheric heat exchange that are true,' I asserted. 'Gore's theory is not coherent with other theories about the Greenhouse Effect. It is acceptably post-modern to fault an inconsistent theory and that is what I am doing. Thank you for raising this, Pamela.'
'Is Gore's proposal true, that renewable energy can be substituted for fossil fuels?' asked Leblanc.
'Untrue, because an enormous cost is involved, beyond the means of energy users and society generally. The standard of living of Australians will be reduced. The poor performance and high cost of the substitute technology is being concealed, but already many energy users are hurting. It is doubtful that society can afford to complete the transition, nor see it through.
'Next reader, please.'

The Net Zero intervention assumes that carbon dioxide can be displaced from energy supply or removed from the atmosphere. Renewable energy can achieve the former but CO2 in the air is a byproduct of many activities and industries and they would shut down if it is banned. There is no easy way of converting it, or sequestering it, into another less polluting form. The only viable solution is to reduce combustion and this means scaling down energy production.

Whereas the Luddites opposed new technologies that took jobs, there could be a new movement of Energy Luddites who oppose energy production. Activists could want to oppose automation, such as lifts, by taking the stairs. They could eschew one-person cars and prefer energy efficient public transport. They could want active transport and bicycle routes. They could reduce printing, paper, packaging and fast food. Homes would be furnished and heated austerely, reverting to standards justified by stoics. Above all, they would be minimalists, making do using a minimum of energy. Believing energy use is anti-social, they would defect from the hamster wheel of material acquisition that causes most energy use and pollution. They would be altruistic in conserving energy for the community to use, leading a campaign against energy greed.

'Energy has rules of chemistry that determine how energy can be extracted, converted, utilised and wasted. This knowledge lies beyond the experience of the politicians wanting to get re-elected by changing energy supply. They rely on experts and are led to consider unlikely schemes, such as 'green hydrogen', that fit the output demands but cannot access suitable energy supplies. Schemes proposed are infeasible and divisive, costly and slow, incapable of rescuing energy supply from falsity.

TASK 32
What are the inputs and outputs for a coal-fired power station?

TASK 32 Answers
Inputs: coal, oxygen, water
Outputs: electrical energy, carbon dioxide, ash

CHAPTER 33 THE ENERGY SPECTACLE

'First reader, please.'

So far we have studied some energy technologies. We have investigated energy use in the home and in managing of diets. We have encountered the major problem of fossil fuels used in power stations, said to be the Greenhouse Effect. We have compared electricity from renewable sources and the supposed reduction in air pollution. We looked at a future with living in space. Our city could be made more liveable by encircling it with an underground railway. Our concern as citizens throughout has been that the government should provide the Australian people with wise and economical energy supply.

The movie An Inconvenient Truth caused close scrutiny of the traditional energy resource: coal. Australia has heaps of the stuff. We thought our electrical energy was secure for the 21St Century, but the movie falsely raised the spectre of pollutant-induced climate change, unless we changed immediately from coal to renewable energy, with dramatic consequences,. Now, 20 years later, renewable energy investment does not look like achieving the goal of Net Zero by 2050 and power stations scheduled to be closed down are being rescued by the nanny state.'

'Is the transition from fossil fuels to renewable energy being managed in the public interest?' asked Leblanc.
'It is expensive and complex restructuring is being paid for by energy users and the wider public,' I said.
'Next reader, please.'

Can we leave energy and climate change in the hands of capitalists and governments, trusting them to provide sensibly for future needs? A 'nanny-state' over-protects people, providing more care than some people want.

A French philosopher, Guy De Bord, in 1967 published his book The Society of the Spectacle. It revealed that the media, the entertainment industries and corporations are engaged in promoting and creating 'spectacles' to profit from media audiences, winning votes for governments. In this view, the appearance of pursuit of mitigation of climate change is sufficient, because the outcome is elusive.

The conundrum is widely familiar to media audiences who pay for entertainment and are exposed to advertising by energy and climate change interests, car manufacturers, travel and holiday purveyors, to cleaning product suppliers. Audiences are inundated with images they can consume and be consumed by, as part of a concerted spectacle.

'How can an audience be consuming and consumed?' asked Biggs.

'Audience members like to eat fast food as part of experience in an immersive audience,' I said. 'An audience for sport consumes fast food together, consuming itself by participating in singing traditions and unifying in a Mexican Wave.'

Next reader please.

When people are caught up in mass excitement, they open their hearts and their wallets to sellers of dreams, goods and services.

The recent FIFA Women's World Cup, created a spectacle that won Australia international prestige, government support, generated large profits from ticket sales, from broadcasting rights, from merchandise sales, from player airfares, from fan tourism and more. The spectacle induced unity in a large international audience, kept together in national and politically motivated groups, consuming and consumed in many ways.

The spectacle audience is largely passive, respecting each other's rights, enjoying belonging to a kindred group, valuing entertainment together, wanting safety protected by the group, obtaining identity they want, commencing friendships, sharing knowledge, united in consumption of fast food and merchandise, discovering companionship and existential happiness.

With the price of electricity escalating, the energy spectacle is watched by electricity consumers and is subsidised by the government, with the leaders milking the situation for dollars and votes. It seems likely that this source of entertainment could continue for years. Only the Greens are impatient for renewable energy to takeover but it would cost more than their government could afford.

TASK 33
What audience does a power station have that can bring profit and votes?

TASK 33 Answer
Electricity consumers, workers and voters.

CHAPTER 34 CLIMATE CHANGE IS A SPECTACLE

'When the nanny state endorsed it, the movie 'An Inconvenient Truth' animated climate change as The Spectacle,' I said.

'Why was the movie's message embraced so fervently?' asked Tracey.
'Global warming propaganda touched a raw nerve that became climate change and magnified public concern,' I said.
'Next reader, please.'

The Spectacle provided images for all the media to publicize. The nanny state wanted to lead a campaign against carbon dioxide. It would bring funds and enjoin the corporations in a campaign that would attract votes.

Climate Change is a foundation of the Spectacle. The Spectacle earlier had images of glaciers melting, floods and cyclones. The update has a conglomeration of technological images, like the Japanese anime' Howl's Moving Castle, an energy fantasy, or the technological spectacle in Alinta Energy's advertisement depicting a mobile conglomeration of renewable energy devices, seemingly deployed as a synthesis, capable of everything. In the advertisement, the spectacle walks on pontoon feet and has an oil derrick, a wind turbine tower, a hydro dam, cooling towers, a furnace stack, a solar panel, a gas turbine, a nuclear reactor and a pumped storage dam. It is an incongruous fantasy, a dreamlike spectacle bringing a solution, to keep the politicians talking, proposing new combinations. People signed up with Alinta to be their provider, bought Alinta share certificates, lured by energy consumption and exotic technologies, in the spirit of the South Sea Bubble of 1720.

The South Sea Company had been at the centre of a complex and tangled web of financial manoeuvrings; deliberately obfuscated dealings around selling shares and government debt; the exploitation of publicity and hype to inflate stock prices; as well as bribery and corruption. Amid frenzied dealing, the shares were in a secret venture 'No-one is to know what it is!' which turned out to be the slave trade. That spectacle was founded on falsity, as is the spectacle today.

Today's energy market is an amazing spectacle, promising a technological revolution, worthy of slack-jawed astonishment, but fatally misleading.

The graphic above depicts a climate change spectacle, a fantasy of technologies, climate change and nanny state images.
Source: By Christos Georghiou (Adobe Stock extended licence)

The climate spectacle put images on screens that captivated a national audience. It had spruikers who told entertaining accounts of how energy could be produced without causing more climate change. The spectacle has won support for government action, but it wasn't clear what if anything would be done. It didn't matter because they believed in the spectacle. All that mattered was that climate change would be opposed.'

'How did the Spectacle take over?' asked Tracey.

'Debord's Spectacle took over in three steps, years or weeks apart.
'STEP 1 is when a community of workers are alienated from their work and unite in ritual worship of their God and fetish objects. There is excitement, collective effervescence, like a fever.
'STEP 2 is when the workers remain separate at work but gather outside work in political action groups, coming together to challenge their wages.
'STEP 3 is when they share spectatorship without class consciousness, refrain from political activity and are without a common enemy capitalist. They absorb the spectacle energy, passive and impotent, becoming 'homo-spectator' who relates to others through images such as social media following and reality TV games. They are doubly alienated, by their work and as spectators.
'Next reader please.'

The Spectacle had government support and people were reassured that something was being done. There was no question of leaving droughted farms, bushfire victims, flooded homes and cyclone disasters to be countered alone, by solitary resilience or by local beneficence. People expected the nanny state to take care of them, doing more than organising donations. The climate was harsh and few were opposed to preventing global warming, although the cause wasn't certain and Australia would be a minor player in global action.

Images have appeared in media, public appearances and news feeds. They have included remedial actions such as closing of coal-fired power stations before the consequences had been considered. Most viewers of the spectacle were passive, absorbing the energy it generated. The structures of their viewing bonded them together, as was intended when the capitalists promoted the spectacle. Capital had accumulated until surplus value had accreted and they could sponsor TV viewing with class consciousness impotent. Workers became spectators.

'I hope I have convinced you that climate change is a spectacle composed of media images,' I said. 'The talk that goes with them is almost entirely false.'
'The spectacle they are promoting seems to be attractive,' said Jessica. 'It seems rather far-fetched but is possibly worth contemplating.'
'Next time when you are being entertained, look to see who is getting your money and your vote,' I said. 'You suffer the three stages, in which you are first a village worker who becomes alienated from your work, joining with others outside work, declining Marxist revolution and becoming a snackfood-scoffing couch potato
'Why do the workers become spectators?' Jessica asked.
Next reader please.

Capitalists generate the spectacle to capture class consciousness, to divert workers from challenging their wages. Debord adopted Marx's concept of worker alienation from the means of production. They share identity as spectators, passively viewing as spectators or fans. They absorb the spectacle energy, impotently, becoming 'homo-spectator' who relates to others through images such as fandoms.
The nanny state's intervention has an image of indulgence. The people sit back being entertained by watching the electricity supply revolution and thankful for the government subsidization of a large increase in electricity charges. The inevitable increase in costs of electricity are borne by governments, who will pay for it by tax increases.

'Why is it called The Spectacle?' asked Emily.
'People passively watch their interests being betrayed.'
Next reader, please.

An audience has a procession of images to watch. It explains what is happening now. Workers are entertained by the procession of Pacific nation prime ministers who without definite evidence of local sea level rise, seek compensation and immigration rights for their people. The watchers contemplate with amusement sources of pollutant gases which could be remediated, such as cows' flatulence. The possibility of feeding seaweed to cows is entertaining and profitable for the media purveying this information. Lies about carbon dioxide are told them for votes and profit. The spectacle has no end of diversions to distract its watchers.

TASK 34
What have been the three stages in takeover of climate change by the Spectacle?

TASK 34 Answers
1. Workers idolize the teachings of a powerful leader.
2. Workers join action groups and political parties to pursue their interests.
3. Workers passively enjoy the spectacle and subscribe to its images.

CHAPTER 35 SPECTACLE SUBSCRIPTION FRENZY

'After the movie 'An Inconvenient Truth', coal and carbon dioxide became the target of government climate action in Australia. The nation was powered mostly by coal-fired power stations which it began to close down, regardless that the nation held a large percentage of global coal reserves and was the World's number two coal exporter. The doomsday spectacle cancelled a significant part of the wealth and prosperity of Australians, who were mostly unaware their high standard of living nurtured on coal royalties could be lost.'

'First reader, please.'

Climate change resided with statisticians and climate forecasters for a short while, until they realised no amount of climate modelling would show the way forward. Definition of climate change was taken over by parasites whose aim was to carve comfortable livings and high influence by magnifying the situation in public perception as a problem. The Net Zero goal could calm anxiety from Gore's fake crisis. It was adopted and appeared in the Paris Agreement.

'Why did climate change become so magnified as a national problem?' asked Jessica. 'Was it because no-one had a solution?'

'The effects of climate change were elaborated as distant prospects: sea level rise; glacier retreat; polar ice disappearance; great barrier reef destruction; ecological damage; and more,' I said.

'They seized images of calamities, with no attention to sampling,' said Leblanc. 'They had potential for profits and votes. There was a pile-on of believers.'

'Read on, please.'

Climate change was all embracing and omnipotent, urged to be mitigated by universal planet wide action. With renewable energy as a panacea, the spectacle had carbon dioxide pollution. Concepts of carbon footprint control and carbon sequestration proliferated with little attention to practicality.

Fossil fuels, which had been the mainstay of the economy, were vilified as pollutants. The government refused to licence coal export, despite there being lucrative markets in developing countries, nor did they allow new coal mines, nor did they seek conservation of energy. Electric cars were added to the spectacle, although more electricity supply was not assured. Airlines resumed after Covid, using fossil fuels with impunity, because no alternative method of propulsion was available, as if jet travel was essential and a right.

The spectacle that had begun in energy and technology spilled over into climate change images, a spectacle of technology and energy that included alternative methods of travel, as if travel was its own reward. Online meetings and zoom conferences had insufficient kudos. Fossil jet fuels were allowed to continue.

The spectacle is the inverted image of society in which relations between climate causes have supplanted relations between people, in which 'passive identification with the spectacle supplants genuine activity' I said. 'Opponents of climate change subscription were labelled deniers, or sceptics and their views cancelled by woke extremists.

'My participation in policy discussions has been cancelled because of my views,' I said. 'These have become talking shops for believers under the aegis of an intolerance of other views. The Spectacle includes a jamboree for climate change scouts and warriors.

'The spectacle is not a collection of images,' Debord wrote in The Society of The Spectacle (1967), 'rather, it is a social relation among people, mediated by images. Accordingly, social groups of believers have gathered with membership validated online.'

'What does The Spectacle do?' asked Kelly.

'It connects people and events with common interests,' I said, 'marshalling superficial unity.'

'Debord's spectacle theory is diffuse and variegated, supported by voluminous argument and observation of variables of psychology, sociology, media presentation, marketing and government actions throughout society,' said Kelly. 'The Spectacle has been fomented by a mob of self-interested persons. They are alienated from their work and are content to go along with social movements out of self-interest.

'What effect will it have on society?'

'These are not the type of people we need to steer Australia away from the climate change morass. In his analysis of the spectacular society, Debord notes that the quality of life is impoverished, with such a lack of authenticity that human perception is affected, and an attendant degradation of knowledge, which in turn hinders critical thought.

'Next reader, please.'

Ordinary people swallow the false posturing of the main players, enabling them to lurch around altering the policy controls, without a public debate about what should be done. The quandary richly rewards its leaders. The spectacle prevents individuals from realizing that the society of the climate spectacle was only a moment in history, one that can be overturned, through revolution that we don't want.

'Debord's spectacle is a dismal prospect,' said Leblanc. 'It sounds as though the people who benefited were the leaders and no-one else.'

'Yes. Debord's theory draws on Marx's socialism at the level of workers' alienation, but stops short of recommending revolution.

'It describes a situation in which people are overwhelmed by conflicting and inauthentic images and sit back, in contemplation and presumable despair.'

'But they are not alone and it could be worse.'

'The spectacle unites them in a fatal garrison mentality.'

'Are the people who watch the spectacle gullible?' asked Kelly.

'Yes. Next reader, please.'

Guy De Bord said: 'The spectacle is a social relationship between people that is mediated by images. In a consumer society, social life is not about living, but about having; the spectacle uses the image to convey what people need and must have. The spectacle pacifies the masses. Everyone participates in keeping it going. The spectacle unites them.'

'Are they like a football crowd who think and act as one, getting carried away with enthusiasm,' said Leblanc, 'expressing their unity in yelling heroes' names or in keeping a Mexican Wave going.'
'Yes,' I said. 'The spread of commodity-images by the mass media, produces waves of enthusiasm for a given product, such as an Iphone, resulting in moments of fervent exaltation like the ecstasies of the convulsions and miracles of the old religious fetishisms. A commodity relationship is created when a trade is reciprocated endlessly, without change of terms. Employment which has been variable, becomes an expectation.
'Next reader, please.'

Audiences of believers hear the prognostications of climate scientists and the promises of technologists that support absurd proposals. The absurdities don't matter to the organisers, because kindnesses of the faithful funds their attendance with kudos.'
So far the spectacle has seemed benign, whereas its role is rapacious and hopeless. I want to persuade you that climate change is a significant part of a 'spectacle', contrived to part viewers from their hard-earned shekels and votes. I want to leave you, having read my narrative, believing that the Spectacle is appearance of a set of images of climate change; the talk that goes with it is false and without hope.

'But instead of pursuing reality the watchers are lulled by images of disasters which affirm their predicament,' said Kelly.

'Lies about carbon dioxide and how the atmosphere is heated are powerful images distorting reality,' I said. 'The spectacle has climate crises, catastrophic weather events such as floods, droughts, coral bleaching and extinction of species.

'Read on.'

Image by image, layer by layer, the movie first suspended and then reworked a new message. The presentation of the movie was emotional and inexorably it brought people to imagine the situation was dire. They were angry the atmosphere was being polluted and they knew they were contributing to the pollution.

In the field of psychology, cognitive dissonance describes the mental discomfort people feel when their beliefs and actions are inconsistent and contradictory, ultimately encouraging some change to align better and reduce this dissonance. They want society leaders to outlaw carbon dioxide.'

'Is accepting the Spectacle like being brainwashed?' asked Emily.

'Yes. The protesters are confident of getting climate action. Government had long ago shouldered a responsibility for climate disasters, the ozone hole, leaded petrol and Chernobyl and they thought they could do it again, with even more success. The government has nurtured a militant following.

Next reader, please.

'The movie fired them up,' I said. 'It took a couple of years for all the academics to climb on the bandwagon, but the mob took the spectacle's bait: the movie - hook line and sinker.

'They really believed they could stop the carbon dioxide and everything would be okay, a deserving cause, like previous pollution scares but magnified in extent. Preventing a substance so innocuous and pervasive was unprecedented.'

'Why wasn't the falsity exposed?' asked Tracey.

'Philosopher Jean Baudrillard's theory is that unless something can be simulated, it cannot become real,' I said. 'The carbon dioxide hoax could not be simulated and so it was hyper-real. Too many university people were on the bandwagon. The universities received money to investigate and were promised more money when they found complications. The universities were in a feeding frenzy and still are.'

TASK 35
How does the spectacle verify that the cause of climate change is carbon dioxide?

TASK 35 Answer
Theoretically but not transparently. The effect couldn't be simulated and was hyper-real.

CHAPTER 36 IS AUSTRALIA A NANNY STATE?

'A Canadian journalist and magazine publisher said Australia is a nanny state.'

'...Australia is becoming the world's dumbest nation...(because of) the removal of personal responsibility and the increase in the number and scope of health and safety laws.' Tyler Brule, 2015

'Is that good or bad?' asked Tracey.
'Bad,' I said. 'He argued that Australian cities are over-sanitised. Many of the laws have been implemented in the expectation that they will reduce violence or improve health and safety. The excessive laws were accused of restricting freedom, ruining livelihoods and small businesses.'
'Next reader, please.'

A 'nanny state' cares for its citizens. If it goes too far, the over-reach is called 'nanny-state' because it over-protects people, providing more than some people want.
He said Australia could be a nanny state, over-protective, interfering unduly with personal choice and unwanted societal control. A nanny state has the appearance of protecting vulnerable people from small dangers, incompetence, foolishness, bullies, abusers and exploiters, like a 'nanny' who takes care of unable, greedy, unruly and innocent children.

'I would like to be taken care of,' said Tracey,' at least, protected.'

'Our governments want to take care of you,' I said. 'Do you know why?'

'To get my vote and my dollar,' said Tracey.

'Next reader, please.'

Our government is self-interested. They want you to let them take care of the climate and pay them for doing it. The Government's response to climate uncertainty has the appearance of trying to protect people from hardship caused by climate change. They alleviate some difficulties but have left personal responsibilities to individuals. We expect to pay for energy and water, with the only control on supply being cost. We don't pay directly for protection against the climate, but we expect relief from droughts, floods and cyclones. The government hands out a lot of money although I am sceptical that victims receive as much as they deserve.

Government in Australia takes care of many things for us. Control over energy, water and housing in Australia is through political democracy, changing at elections from conservative to socialist governments. There is no best system of control, because conditions vary from place to place and time to time.

'Almost everybody is on a state pension of some sort, without money they need,' said Biggs. 'I get nothing from the state despite having paid all of their damned taxes all my goddam life. Whatever happened to the idea of saving for a rainy day?'

'How should the government decide who to help with money and services?'

'They should help old people, ill and disabled,' said Biggs.

'What about the poor?' asked Emily.

'They should have a means test, to see if they are capable of working,' said Tracey.

'It is not easy to decide who to help, because some may be capable of helping themselves more than they want to.

'Next reader, please.'

I want to expose nanny state overreach in Australia that diminishes personal responsibility, undermining self-control. When an authority treats competent adults like children, they learn helplessness and become dependent on the state, losing ability to take care of themselves. The state can become over-protective and authoritarian, interfering in people's lives.

Rousseau's social contract required all people to act for the public good,' read Kelly. 'Soviet communism prescribed state control of religion, health, education, employment, manufacturing, commerce and election of leaders. When the people withdrew their support from the 'nanny', the state failed.

Singapore is reputed to have many more regulations and restrictions on citizens' lives than in other countries. Germany was found to be least restricted of 30 European countries, in a recent survey of regulation of alcohol, tobacco, food and vaping.

'Maybe Germans are least well off,' said Jessica. 'They are without beneficial state services.'

'No,' said Kelly. 'Germans have more free choice. Our governments have legislated to control thousands of products and situations unnecessarily.'

'I don't accept that Australia has to be a climate-nanny,' I said. 'People should take care of themselves if they can.

'It depends on whether you see State actions as under-protective, protective or over-protective.

'State actions could include modifying houses to counter warming, flooding and storms. It would be expensive. Tax payers not receiving remediation could complain of nannyism.

'A nanny state depends on the people's good will supporting those who are disadvantaged.'

TASK 36

What are examples of nanny state over provision in Australia?

TASK 36 Answer
Over-provision for safety, disadvantage and climate change.

CHAPTER 37 WHY NOT A NANNY STATE?

'First reader, please.'

Climate changes could be state-wide but not everyone would want nanny state protection. Over-dependence on the state is possible.

Being nannified doesn't sit well with Australians who foster an image of rugged independence. Libertarians, especially, cite over-protection already.

We're steeped in nanny state laws. We have mandatory bicycle helmet laws, gun control laws, prohibitions on alcohol in public places, plain packaging for cigarettes, pub and club lockout laws and permits for picnics on a beach. Some are ridiculous. These matters used to be for individuals to control voluntarily. Australia's criminal legislation has gone too far.

'Our gun control laws are reasonable,' I said. 'Other nations envy us.'

'I'm not sure,' said Jessica. 'A nanny state excessively controls, monitors, protects, or interferes with people's private actions or behaviours that are deemed unhealthy or unsafe. Having one's domicile identified as a nanny domain could be interpreted pejoratively, as an admission of disability or weak resolve.'

'What is state-like about a nanny state?' Tracey asked.

'The term could echo 'nation state', which is a body of related people in a country. A nanny state has a nanny figure parodying a monarch. The government could be autocratic and would be resented by the people.'

'Next reader, please.'

Utopians like George Orwell have satirized cradle-to-grave care by the state,' I said. 'Scandinavian welfare comes closest to that. Israel, Cuba and former Soviet countries have achieved some success, but opinions about this differ.

Some people want the state to supply everything, with little or no personal expense!' said Leblanc. 'They dream of being securely employed under good conditions, without having to compete with others. To them, being without luxuries would not matter, because everyone would be without them. But equality has never been achieved.

Orwell, in his book 1984, satirised a totalitarian hell, with state control of every aspect of Prole life, including consorting, thinking and talking.

'What is a Prole?' asked Norman.

'Proletarians are the masses, leading impoverished lives performing meaningless tasks, alienated from their work, controlled ruthlessly like slaves by Big Brother,' I said.

'Australia has had totalitarian tendencies. During the Covid pandemic, various technologies were proposed to be mandated: quarantine, masking, vaccination and social distancing,' said Leblanc. 'These conditions of the nanny state were protested in some communities.'

'Those who spoke out during the pandemic were deemed to have unacceptable viewpoints and were ostracized, boycotted or shunned. It was cancel culture,' I said. 'Most interventions were adopted democratically and objectors were usually a minority. Too bad if you were one of the minority.'

'Objectors to nanny state provision are sometimes labelled as 'haves', capitalists, authoritarians or conservatives,' said Leblanc. 'Nanny state supporters are labelled socialists and sometimes derided as 'have-nots' and free-loaders.'

'Babying adults is self-defeating,' I said. 'The baby gets thrown out with the bath water. Provisions intended to keep streets safe for children, sometimes have the opposite effect. Zebra crossings and traffic calming obstacles encourage pedestrian mindlessness. Children don't learn to cross the road safely. They step out into a zebra-fied trap without looking and are squashed. Hoons and rat-runners speed on any streets without speed bumps.'

'Destitute people expect a nanny state in a civilised society to provide the necessities of life,' said Jessica.
'Next reader, please.'

'Poor people want a nanny state to subsidise their rents in a housing shortage, as if the market is incapable of charging them a fair price.' I said. 'They are like babies, fed by umbilical cord. They cling to benefits that others have had to work to achieve.'

'Electricity users expect the government to pay increases, instead of reducing their bills by conserving power,' said Kelly.
'Next reader, please.'

'Our nanny state sometimes helps with the cost of disabilities and this is just,' I said. 'If a claimant gets away with faking symptoms, others could have to provide support. I had a student who claimed to have dyslexia and she was required to write her exams on dark purple paper with a black pen. To mark it I had to be able to read it in bright sunlight. I was ruining my own vision catering to her faked needs, until

I realised her ruse was a rebellion. I refused to take part. When a nanny state wants to discriminate positively, the validity of the claim and its provision can be a significant cost to the community and is sometimes excessive. Forced to comply and write on white paper, her performance was acceptable.

'Next reader, please.'

'Over-protection is rife,' said Biggs. 'A sign displayed at the entrance to a local park reads 'Beware falling branches.' People should not be put in a defensive posture where they go to experience natural conditions. In my opinion, this is overreach, without sufficient benefit to justify intrusion.'

'What's the worst a nanny state can do?' asked Leblanc.

'Promote a person's position at work unfairly,' I said. 'Well-meaning people have responded to inequality with affirmative action or positive discrimination,' I said. 'Their action is controlling and controversial.'

'A nanny's pampering can have a bad effect' said Leblanc. 'Provision of welfare benefits can make the recipient dependent, or even addicted. To avoid creating dependency, the nanny has to consider doing nothing at all.'

'A nanny-state can be called unjust, by virtue of more provision, more identities to recognise and more grounds for controversy,' said Sophie. 'When recipients are designated for benefit, the state can be at fault from 'woke' critics alert to racial prejudice and discrimination. Beginning in the 2010s, an inclusive viewpoint objected to prejudices against social inequalities, such as racial injustice, sexism, and denial of LGBT. Every new category of 'Alphabet' people has new possibilities to confound a nanny state with claims of discrimination.

'Some indigenous people in Australia benefitted from affirmative action at first but it has now been been revoked because it caused resentment in the general community,' I said. 'It is not always possible to establish true equality and some indigenous people are still wanting reparations to be awarded by the legislature for settler invasion 200 years ago.'

'Thank you, Sophie. There can be differences between community groups expectations of nanny state provision and it can be difficult to agree when to apply it.'

'Next we'll review the new demands being made on the state to invest in energy and climate change, then we'll consider whether Australia really needs a nanny state.'

TASK 37
What problems can result from a nanny state?

TASK 37 Answer
Over-dependence.

CHAPTER 38 CLIMATE IMAGES ATTRACT NANNY INVESTMENT

'Below is a list of 16 different ways authorities can supply energy in compliance with Net Zero conditions, usually by the nanny state investing in new energy supplies and facilities. They are proposals bandied around the climate change spectacle, for governments and corporations to invest in, for profit or for votes. Because opinions of nanyism vary, the attractiveness of nanny intervention has to be considered case by case and that is what is done in this chapter.

'First reader, please.'

Commentators without engineering training have put forward proposals to prevent carbon dioxide being produced or entering the atmosphere, as if matter can be 'made to disappear' by paying for research by throwing money at the problem.

'Matter cannot be 'made to disappear' and therefore It is not an acceptable use of nanny-state funds,' said Leblanc.
'Next reader, please.'

Carbon dioxide release is the cost of utilising coal's energy. Burying carbon dioxide by sequestration underground has seemed possible, until the practical difficulties are faced. It could be better not to dig up the coal in the first place.

The climate change spectacle has images that attract nanny state consumption, with wide sympathy for victims. For example, residents whose homes have been flooded, can seek resilient restoration, or even have their homes bought back. Faced with the likelihood of climate change, mitigation projects are virtually unlimited.

'Some flood claim compensation is fairer than others. I'm glad I don't have to draw the line,' said Tracey. 'Perhaps politicians do something useful after all.'
'Next reader, please.'

Technical problems are worsened by deception. There have been many scams involving energy production which contradicted the laws of thermodynamics.

'There is no easy way to disappear carbon dioxide molecules,' said Kelly.

'We'll read down the list, discussing and questioning,' I said. 'Your task will be to show you understand the role of the 'nanny-state' in each situation.'
'Next reader, please.'

1. Relocation of the Tuvalu islanders is a vote winner because relocating them to mainland Australia would be spectacular. A nanny response could take responsibility for Australian polluters said to have caused part flooding of their sand island.

'What is the role of the nanny state here?' I asked.

'The Australian nanny state could want to pay damages to Tuvalu islanders who claim compensation,' said Leblanc.
'It has an appeal to hubris,' I said.

'Continue reading, please.'

2. The nanny state may be needed to create the infrastructure needed to operate electric cars,
There could be spectacular owner benefits and encourage transition to EVs.'

The nanny state could subsidise charging stations.

3. Subsidisation of solar panels could be less than was previously spent on generation and transmission. But the investment is spectacular for replacing public power stations with private solar panels, deemed to pollute the atmosphere less.

The nanny state could privatise domestic electricity provision.

4. Home solar batteries can be subsidised to spectacularly displace reticulated electricity supplies.

The nanny state could subsidise private storage of electricity.

5 New wind turbines are expensive and unlike solar panels, they aren't normally private. They will be subsidised for spectacular investment by communities and by the nanny state.

The nanny state could devolve electricity supply to communities.

6. Time of day metering systems and tariff structures can help match demand and supply of electricity, possibly needing subsidies.

The nanny state could regulate tariffs with subsidies.

7. Burying plant material on the farm can improve soil fertility and take CO_2 out of the atmosphere. Subsidised mulch crops could sequester carbon.

There could be a short term benefit paid for by subsidy.

8. The massive Snowy Mountains 2 project is a spectacular way to generate public electricity with a subsidised dam to recirculate water, using cheap electricity to recycle expensive water.

Electricity generation would be subsidised.

9. Trees take out CO_2 and they can be conserved by planting trees and protecting forests and bushes from clearing and fires that produce carbon dioxide with subsidies.

Planting and clearing of trees could be subsidised by the nanny state.

10. Energy use, emissions and waste heat can be reduced by subsidising equipment and resources.

The nanny state could oversee the installation of energy using technologies.

11. There can be economic growth without growth in energy consumption. Other functions of the economy can be tracked, such as growth of leisure facilities.

The nanny state could subsidise low-energy ventures.

12. Investment in cattle breeds and feeds could reduce methane output significantly, to improve local and foreign climates. Methane could possibly be sold.

Transition to less gassy breeds and feeds could be subsidised by the nanny state.

13. Greenhouse gases are produced by landfills and their reduction could be worth subsidising. Methane could possibly be sold.

Gases production could be reduced by nanny state regulation of wastes in landfills..

14. Preventing leaks of refrigerants and substitution of ozone friendly refrigerants could be worthwhile.

Regulation by the nanny state of refrigerants could prevent ozone catalysts escaping.

15. Tree planting could mitigate under nanny state control some of the global warming from CO_2 in aircraft exhaust gases.

Aircraft operators using fossil fuels could be required by the nanny state to plant trees.

16. Australia's coastline could be reserved for thousands of sky-scraper-sized wind turbines.

Bird watchers, fishers, shipping companies and recreational users of coastal waters could be compensated for loss of amenity.

'These projects are for a nanny state with a bloated budget, for aggressive investors to intervene in the lucrative climate change market,' I said. 'Many other spectacular climate change investments are possible.'

'Many of those projects are nonsense, in my opinion,' said Leblanc. 'How was the list compiled?'

'These were the best-supported projects proposed by the government, opposition and in the media,' I said.

'Next reader, please.'

The purpose of the listing above is to reveal 'climate change' images which are part of the Spectacle touted for government or for private support. The 'nanny' term likens such a government to a nanny who is rearing children: cosseting individuals who are not yet mature. It is generally supposed that entrepreneurs will not take on the challenge. The actions proposed are like a religion, devoted to stopping climate change, under illusions that they will help people live. The climate change spectacle overreaches and is intrusive.

'Who are the cosseted individuals?' asked Emily.

'They could be workers and young people engaged with the spectacle, but not investors, who would be standing back,' I said.

'Next reader, please.'

The actions listed above are predicated on carbon dioxide causing climate change. The view I have stated is that the effect is small and the Net Zero policy is reckless. The spectacle includes climate change and there are many opportunities for profit and votes. There would be no harm in this except that spectacular images of climate change are often false, with expensive taxes. People could lose their income, amenities, property, homes and jobs.

They say that after the transition, we will be better off. The Net Zero intervention is supposed to end all hardship.

'What is the Net Zero intervention?' asked Emily.

The Net Zero intervention is part of The Spectacle,' I said. 'It is cutting of carbon gas emissions to a small amount of residual emissions that can be absorbed and durably stored by nature and other carbon dioxide removal measures, leaving no more in the atmosphere than is being removed,' I said neutrally, trying not to sound sceptical.

Although some of these proposals could be achieved in Australia by the Net Zero initiative by 2050, the reality within most nations is that emissions are increasing. The climate change spectacle of Net Zero has a dreamlike quality. Investors could be burned when it is not achieved.

'Few of these investments will materialize and the nanny-state will be committed to carrying dozens of lame-duck projects.
'Nanny state investment seems haphazard,' said Emily. 'There is no unified commitment and participation in Net Zero does not seem to concerted by the nanny state or by anyone.

TASK 38
What is nanny-state about Australia's subsiding of early childhood education?

TASK 38
Some people could regard it as over-protection.

CHAPTER 39 DOES AUSTRALIA NEED A NANNY STATE?

'Opportunities for the Australian state to intervene in its citizens' affairs is a moral choice from individual belief, but more often is imperative for a person's lifestyle to continue,' I said. 'People are required to conform. For example, subsidised jet travel. There are so many avenues for state nannyism to intervene beneficially that many people want a nanny state to be there for them at all costs. For others who value independence, loss of individual freedom to decide matters would not be offset by sufficient benefits. Fortunately, people are not asked to make a choice once and for all.

'The perspective here will be to indicate areas of public policy in which individuals could benefit and others where they might not, particularly those involving energy policies.

'Australia's governments have enacted legislation to supply citizens with energy, water and other utilities. They can protect citizens from rapacious suppliers and effects of climate change. They can protect people by reducing violence and improving health and safety. If they are over-zealous, they can interfere, reducing private choices and commercial freedoms, controlling society.

'Many of the laws have been implemented in the expectation that they will reduce violence or improve health and safety. The excessive laws are sometimes accused of restricting freedom, ruining livelihoods and small businesses. Australia could be a nanny state, over-protective, interfering unduly with personal choice and unwanted societal control. It could protect vulnerable people from exploitation, hardship and dangers. Some people could feel over-protected with more provision than they want.

'Most people expect the government to lead the way, passing laws to protect them. Take pollution, for example. Filled with hubris after controlling DDT, Australia's environment protection agencies have vigorously striven to control various types of pollution. They have struggled with chlorofluorocarbons, smoking, lead in petrol, asbestos and refrigerants. Now they are trying to rise to the challenge of carbon dioxide. By supporting the Net Zero campaign, they may have bitten off more than they can chew.'

'Governments want to fix societal problems,' said Jessica. 'There are bound to be risks.'
'If it doesn't work, they will say, 'too bad we'll try something else,' I replied. 'What they can't do is sit on their hands, hoping a solution will pop-up.'
'Why not?' asked Buck.
'When they win, they get kudos. Although losers can be held to account, they usually have to absorb the loss privately.
'What if an industry is out of pocket?' asked Kelly.
'The government will say 'We'll pay compensation''
'Stopping coal exports does away with the nation's main source of revenue,' Leblanc said. 'A lot of people will be out of pocket.'
'They sold off the state power stations to private companies, conditional on being able to operate them and recoup the price they paid,' I said.
'Now they're having second thoughts,' Kelly said. 'They're going to have to renegotiate.'
'The greens want blood. All they care about is shutting down the old technology,' said Jessica. 'But electricity consumers want a reliable supply at as little extra cost as possible, which is diametrically opposed.'
'The potential for compensation for climate change is bottomless,' I said.
'Next reader, please.'

The nanny state could be expected to compensate people who claim their living places are threatened by climate induced change, such as rising sea levels and habitat destruction. Dumping waste into the air hasn't come up before, because attention has been on trace chemicals and corrosive gases and the other substances so far regarded as benign. Now government is emboldened to act on climate outcomes and the weight of the law could swing behind the Greens. The nanny state could have to compensate people.

'Climate outcomes are like the weather,' said Jessica. 'How can the nanny state be held liable for that?'
'Now that it is accepted that certain activities like fossil fuel use are controlling the weather, every man and his dog will be claiming compensation!' I said.
'Next reader, please.'

'Australia's weather conditions are quite predictable, allowing governments to protect individuals more than in some other countries. Should private risk be covered by the State? What could be the danger from over-protection? Would lame businesses not have the resilience to recover? Could businesses depend on the state for funds?'

'Businesses are usually reluctant to seek nanny state favours because they usually deal with regulation, taxes and getting approvals,' said Leblanc. 'If the state gives with one hand, it is likely to take back with the other.'

'In the outback, there is often hardship,' said Norman. 'Does Australia need a nanny state to mitigate the harsh conditions? I don't think the 90% of the population who live in cities will want to support the 10% who live in the outback. It is a bottomless pit.'

'The new outback citizens are likely to be well-heeled with an air conditioned utes, telephones and digital services. To access medical, *dentistry, welfare and recreation amenities available in cities, they will travel and retire early.*

'Losers, even in a small way, are not likely to pass up opportunities to claim disadvantage.'
'Read on, please.'

Envy is not usually far away when nannying is criticised. Handing out of money, goods or services, for recipients to use without obligation, can be regarded as 'free-loading', or unearned benefit e.g. compensation for disaster. When the state distributes windfall benefits like these, nanny-statism can be derided by anyone envious.'
Haves may not want have-nots to receive benefits they have paid for e.g. free or subsidised tickets, access to facilities, services or goods.
Australia's health system tries to afford equal treatment for everyone. Individuals choose between public and private health services, based on their preferences and needs. The nanny state pays for expensive medical treatments for people without insurance.
'The nanny state gives employees 8 days of paid public holidays, and in addition annual leave of a minimum of 20 days per year.
Australia has large distances with sparse and underprivileged communities in the interior. Inequalities of location, between city and outback, are difficult for private businesses to supply equally. Subsidised nanny state services are expected to provide equilibration, rather than forcing internal migration.

'What is the main disadvantage of nanny care to Australian society?' asked Leblanc.
'I am reconciled to having nanny state protection where this does not create dependence on the state,' I said. 'Australians are encouraged to take responsibility for their lives at every age. If nanny state over-provision is causing welfare dependence, it is an unacceptable drain on public resources.
For example, Australian governments are massively embroiled in trying to lead public opinion to accept a dubious attempt to protect

individuals from effects of climate change. I have heard that people are building beach houses in the expectation they will be compensated for sea level rise. In my opinion, it is overreach. We do need a nanny state, but too much is harmful.'

TASK 39
Does Australia have any local conditions that justify a nanny state?

TASK 39
Australia has climate challenges, isolation of individuals, lack of community, lack of self-care.

CHAPTER 40 A SPECTACLE OPPOSING ENERGY MINIMALISM

'First reader, please.'

I have justified nanny state support of our remote population and proposed nanny help for citizens who live at isolated locations and encounter disastrous weather conditions.

'Surely, people who choose to live in such places should accept the consequences?' asked Biggs.
'I don't think the state can be expected to relieve all eventualities,' said Kelly. 'They could aim to remove unusual hardship.'
'That seems reasonable,' I said. 'We might expect that Australian government provision for hardship would be conservative and even minimal, assuming local people would respond stoically. But their responses are demanding.

'Read on please.'

Australia for many years was awash with cheap petrol pumps and discounted domestic electrical appliances. As energy prices have increased, cars and washing machines have become more expensive, although more energy efficient.

The images become autonomous and represent other images than spectacles. There is no reality; everything is subject to change. These spectacles have images that keep things the way they were. Their appearance of saving the world is what counts. Faking it is making it.

The images are curated to commodify people's emotions. There is jealousy, resentment, stigma and rejection. Hardship is a spectacle needing provision of energy, water, food and other necessities. It has images of standard and sub-standard living.

Appearing is more important than being. The technology is an elaborate spectacle, degrading life. The spectacle is the heart of our heartless society. In other words, it becomes the centre of our reality and, thereby, it makes our reality into an unreality.

'Where does this spectacle happen?' asked Norman.

'Everywhere and nowhere,' I said. 'The spectacle can rule at work and can get taken home and out to social venues, where the images control entertainment, shopping and even family events. Australians have accepted replacement of electricity generation by more expensive supplies presented as part of a climate change spectacle of abundant renewable energy. This would not be presented for consideration, but as images encouraging them to use energy and pay whatever price the supplier or government charges.'

'Believing the nanny state will pay for spectacle energy costs is deceptive, because the community pays in the end,' said Leblanc. 'Money will come to the spectacle from the nanny state.'

'De Bord explains that energy and climate change are merely appearances and images.,' I said. 'The spectacle is the prospect that faces people and that the state helps pay for.

'Next reader, please.'

Insofar as we now take representations to be more truthful than concrete reality, insofar as we hold up spectacular images as authorities, we are blind to the real truth of our social reality. The spectacle is capital accumulated to the point where it becomes image.

The consumer spectacle is the most efficient way to pacify the working class and to make workers shut the fuck up about their shitty lives of exploitation and alienation. De Bord observed imminent revolt by the workers and his theory avoided another revolution.

Energy has become a spectacle of images detached from every aspect of life, merging into a common stream with climate change and the former unity of life lost forever. Reality unfolds in a new generality as a pseudo-world apart, solely as an object of contemplation.'
The Society of the Spectacle, Guy De Bord 1967.

'The spectacle is a pacifier that keeps the proles quiet,' I said.
'Renewable energy is being presented as a bonanza, as a false spectacle, ignoring that energy costs will greatly increase and supply may possibly fail in bad weather,' said Norman. 'Consumers feel a chill wind of failing supply and they purchase with caution previously unknown to them.
'Next reader, please.'

The spectacle of renewable energy shortfall and even black-outs has been suppressed. To avoid disaster, the electricity authority ought to adopt minimalism in all its operations, but it seldom does, because that would reduce revenues.'
People have accumulated private capital and commodified energy supply, with their participation limited to securing supply obligations and rebates from the nanny-state for cost over-runs. They are without supplier or brand loyalties. The supplier they are connected to has nothing more to offer than they could get from other suppliers, who routinely try to poach their business away. Amid publicity, a consumer is offered a discount but when the small print is read the cost will be the same as from competitors. Energy is a commodity, sold without allegiance or tradition having any value.
The advantages of being supplied with energy as a commodity is predictability, with the price known in advance. Changing to a different supplier is possible. Government interventions could control energy prices, but it destabilises market participation. Governments tampering with energy prices has more to do with gaining voter support than securing better terms for customers.
The images are designed to seduce us, lacking efficiency and economy. There is risk of economic collapse. The spectacle has images

of extravagant lifestyles, encouraging customers to discard goods that are still useful and consume wasteful amounts of new products.

The value of flexibility to the community from moderated electricity demands and minimal provision is neglected. Electricity suppliers should cultivate minimal demands rather than gouging maximum sales of installations and electrical energy. The public needs encouragement to minimize peak demands and conserve electricity, rather than to fill suppliers' coffers. With a minimal spectacle, consumption of electricity can transpose into a more critical and resilient society.

The added benefit of minimalism is to promote conscious decision-making about belongings, time, energy, and relationships. Minimalists are authentic, valuing simplicity and rejecting superficial mass consumerism. They aim to highlight the beauty, essence, and true purpose of things in their lives. When you call a person a minimalist, you're describing their interest in keeping things very simple. A minimalist prefers the minimum amount or degree of something.

The minimal amount varies from home to home but the Spectacle has only one level of provision, for items people need help to afford. Establishing minimums is dangerous. A minimum can be like a wolf in sheep's clothing. Some people are more able to go without than others. Minimalism opposes the kind of capitalism espoused by Adam Smith when he extolled the virtues of the baker who supplied his community with bread. Supplying electricity to affluent consumers is less virtuous than supplying bread to the hungry. Promoting electricity supply to a community served by weather-prone technologies at full capacity is more precarious and more expensive. The lifestyle is less disciplined, requiring more patience and with less peace of mind.

'It seems we need minimalism but suppliers want to oppose it,' said Tracey. 'Isn't there anything we can do?'
'We can save energy,' I said.

TASK 40
How can individuals minimize their energy use.

TASK 40 Answers
1. Resist supply by the spectacle of Big Energy.
2. Minimise peak demands
3. Conserve electricity.
4. Prepare for blackout.

CHAPTER 41 SAVING YOUR ENERGY

I welcomed Councillor Sparks to our online podium.

'Good day, students,' he said. 'The news is not good. The Electricity Supply Committee has increased energy prices to three times their level yesterday. In the Town Hall we have had an emergency meeting to consider how people should respond. Your teacher has invited me to advise you what to do. When you go home you can explain our situation to your families.'

'We have been waiting for the Council to fund installation of solar panels. Unfortunately, the panels haven't arrived yet. Their electricity would be cheaper than the coal-fired electricity we have been getting.'

'Pig's ass,' said Buck to Leblanc. 'Solar electricity funded by borrowing is more expensive than from old coal-fired stations with their capital paid up.'

Councillor Sparks continued. 'We cannot afford to get it wrong. We are not absolutely sure of the cause of climate change, but we cannot risk not investing in expensive renewable energy. It is the only alternative available. The old polluting powerhouses are finished, Comrades, we don't have a choice. We'll have to go with the solar panels and hope for the best. But they won't be installed until next year and in the meantime we're going to have to ration the power supply and electricity will cost three times the recent tariff rate.'

I had seen this coming when they shut down the power station before the solar panels were installed. It had seemed like a recipe for the disaster we now had.

'Climate change could be caused by overuse of energy,' said Sparks. 'After it has been used, the waste heat is at too low a temperature to be used for anything else. Most goes into the oceans, or space, but the increase in ambient temperatures they have measured could be caused by spent energy. Instead of trying to reduce climate warming with solar panels it is logical to stop wasting heat.'

'He hasn't admitted the failure in the supply of panels, when that is the obvious problem,' said Jessica.

'An individual can reduce their energy use by more efficient living, benefiting the community and gaining a sustainable future,' Sparks said.
It was an attempt to blame us and we muttered resentfully.
'We want a reduction in all forms of energy use in our district: less heat; less movement; less fuels; and less radiation,' Sparks said.
'Will we have less climate warming locally?' I asked belligerently, as if local access to global benefits from Australia's Net Zero plan might not be forthcoming.
'Atmospheric warming could be reduced far afield and there will also be an immediate cooling effect here,' he said.
I didn't believe it.

'Would that be from trapping less heat locally?' I called out to Sparks.
'No,' he said. 'Renewable energy is not expected to reduce warming locally except by holding down the global average. I'll stop now for a short break.'
I began chatting with the students.
'I don't like where he's going with this,' Kelly whispered.
'I like humans cutting back energy, but we are at school and shouldn't have to cut back,' Jessica said. 'We hardly use any energy. It is essential for our growth.''

'Energy use is a natural part of life,' said Norman mournfully. 'We use energy to grow food, we use food to gain energy and we use energy to do work. Even when we are sleeping, we use energy. To reduce energy is to reduce living. How can we cut back?'

'I agree,' said Biggs. 'We students cannot reduce energy consumption. We are already at a bare minimum and it could mean starving.'

'Shh,' said Jessica.

While we waited, Buck was beating a bin like a drum and chanting: 'Fe Fi Fo Fum! We can smell extortion!'

Some of the others stepped in time to his beat, forward, back, forward, left, right, right, left, joining in his chant.

Sparks came back.

'Comrades,' he said, his tone severe. 'Only essential energy can be used in our community from now on.'

'All energy ends up as heat in the environment, as global warming, so we must use no more than is necessary. Meat eating will end, because it is energy intensive and infringes on animals' rights, without benefitting human nutrition.'

'Whoopee-do,' yelled Emily. She was already a vegetarian.

The students applauded loudly and chattered noisily. Vegetarianism had always seemed out of reach. Sparks waited for quiet.

'Each of you will get a personal energy quota,' said Sparks. 'You will receive ration cards to collect basic unprocessed foods from food stores, like potatoes and vegetables.'

'Shit,' said Buck.

'At least we'll be getting some energy,' said Jessica.

Sparks sipped from a glass of water.

'I will assign each of you with a quota for your energy use that you must not exceed. There is an app you can use to monitor your usage. You will be allowed one return ride each day on the council bus. Anyone who fritters away their quota will be without transport. Adult car owners will be able to get fuel for one return car journey each day, provided they carry at least two passengers.

'You must conserve your energy,' said Sparks. 'Running is not allowed because it is inefficient, except as part of organised sport.

'Energy use of all types has to decrease, especially food consumption. The right thing to do is go without or use something else. Instead of electric light, you should read in daylight. Shop only once per week, share car journeys with others and do all your stops in one trip. Park vehicles on a slope facing downhill, for easy vehicle restarting. Get the most useful work from every drop of petrol and every joule of food.

'You will follow these rules all day every day,' he said, pausing for another sip of water.

'There isn't enough electricity for heaters, nor for air conditioners, nor for electric motors. If you are cold, close windows and doors, or wear something warm.'

'Are you serious?' said Kelly. 'Are you putting us on?'

Sparks spoke forcefully. 'No, this sets your lifestyle, for us all to survive, from now on.'

The animals' fear was silent. Pamela began sobbing hysterically, until her companions stopped her.

'Grow hair to insulate yourself,' Sparks said. 'You must eat little and move hardly at all. When moving is necessary, you must proceed at a slow and even pace, by the most direct route, which is normally in a straight line.'

'It's stupid,' murmured Norman. 'The best way to insulate me is with a decent amount of food to put on some fat.'

'Shh,' said Tracey.

Sparks spoke more quickly now.

'Electricity for pumping of water must be minimised by reducing water consumption. Hot water from boiling eggs and potatoes is to be reused to make tea, or for washing and hot water bottles. Adults can bathe for one minute every day. Children can have a cold shower lasting under 2 minutes once per week if an application is received in writing. Water from washing and showers will be reused to irrigate our gardens.

'Our essential nutrients will be high yielding vegetables, requiring a minimum of work to grow and easy to prepare, like potatoes. Without diesel, tractor work will be replaced by horses and oxen, doing the ploughing, cultivating and hay making. You can bring back into use the old machines stored in your sheds.

'Electric motors will cease to operate lifts and escalators. Stationary power will come from water wheels, treadmills, horse mills and steam engines as in olden days.

'Our horticulture will grow varieties requiring little manure, little water, no pesticides, cultivated by animal energy, without tractor energy,' said Sparks. 'When Cuba's oil was blockaded, this is how they survived. Potatoes, grains, corn and rice take least energy. Pumpkins, corn, lettuce and root vegetables require moderate energy to tend and are permitted to be grown. Strawberries, tomatoes, beans, radishes and tulips require intensive energy and are forbidden. Cut flowers are not permitted.'

Leblanc raised his arm.

'Who will benefit from all our sacrifices?' he asked Sparks.

'People starving in foreign countries on the other side of the world. They will be able to enjoy familiar climate conditions, indulging in their traditional activities. The value of our adherence to the Net Zero policy is that climate change will stop here.'

'Can you be certain of that?' I asked.

'If everyone works together, we can survive climate change,' said Sparks. 'I've finished what I came here to tell you. I hope our plan seems reasonable and I hope you will play your part. Thank you for your interest.'

He shook my hand and left.

'Is he part of the spectacle?' asked Kelly. 'An energy policeman?'

'Yes,' I said. 'Over the top.'

'What if we continue as we were?'

'Foreigners will be suffering from climate change that we could have opposed,' I said.

'Would they know?'

'Gore said it would be inconvenient,' said Emily. 'It will be.'

'It's a spectacle we can contemplate,' I said. 'Reality will be different.'

TASK 41
How are they required to save energy?

TASK 41 Answers
1. Accept rations and travel restrictions.
2. Conserve electricity and energy.
3. Grow own food by manual labour.

CHAPTER 42 TRENDING

'The purpose in this last chapter is to sum up availability of energy and water supplies in Australia and any trends in provision likely to affect us soon,' I said. 'Certain campaigns are underway and commitments are being made. Current trends could indicate what to expect.

'First reader, please.'

Energy supply has kowtowed to a spectacle of climate change by minimising use of fossil fuels previously banned. Energy supply contracts are being renegotiated and installation of new technologies is uncertain. Substitution of solar panels and wind turbines for coal-fired power stations has been deemed necessary to meet the Net Zero obligation. It is already behind schedule and could fail to be realised.

Nobody knows what performance to expect from renewable energy. The suppliers merely know the deals they have cut and how they can fake performance. If they are found out, they have other attractive deals they can switch to.

Observing from a distance the feeding frenzy for energy supply, proceedings in our community are like a corroboree held for the supplier clans and their friends, where they indicate to guests what government investment they want. The leaders are secretive, but there are strongmen who share out the action between the local energy barons, who jostle for a piece of the action, with many threats and occasional violence.

'Are you complaining about what is going on in business and government circles?' asked Leblanc.

'Yes, I am,' I said. 'I don't know exactly what is going on, but it seems to me that ordinary energy users are being screwed over. The *capitalists' method is to select a piece of the energy supply system, a contract, land, a right of way or a power station. They outline a plan to obtain funding from the government for promises, quid pro quo, or for cash, or a contract. If their shareholders would be sufficiently rewarded, they buy into a play themselves. Energy market revenues are huge and lucrative, with only a veneer of government supervision. If they can get a foothold, their legal advisers secure their positions and then wriggle out obligation free.*

'It's bad,' said Kelly.
'It could be worse,' I said.
'Next reader, please.'

Pieces of the supply system are being sold by the government secretly for a song, then redeveloped for a different use. The government's interest is to acquire a slush fund to apply to the next election. The government controls developments, benefitting its interests, applying its authority to give a semblance of control.

Australia's international commitment has included coal export restrictions and closing down of domestic CO_2 polluters. In the meantime, global temperatures have continued to increase erratically and slowly. Many new coal-fired stations have been built in China, India, Africa and other developing countries. Australia is the major coal exporter to nations who rely on coal for electricity supply. The increase in atmospheric CO_2 has abated only a little. Since 1982, the amount of warming during 40 years has been only 0.8°C total. Any reduction might not be visible for many years.

'I know you don't believe it, but a lot of people approve of Australia's banning of coal combustion,' said Leblanc. 'They have all their hopes for climate control pinned on ending fossil fuel use.'

'The sooner they realise that we have a future with fossil fuels, the better,' I said. 'There have been few closures of CO_2 producing plants in Australia and worldwide. The gas is only 0.042% of the World's atmosphere and keeps the World warm, like the other gases. CO_2

could be just another greenhouse gas capturing solar infrared and using it to warm the Earth's surface and air.'

'Next reader, please.'

Conservation of heat energy from all sources, including spent heat from automobiles and electrical devices, could go a long way to reducing global warming. Atmospheric warming probably results from industrial processes and economic growth. To reduce warming, countries could turn attention from industrial growth to social development. Prole art, fashion, hair, music and sport can create employment and wealth, without much warming of the environment.

'Can the proles become liberated by overthrowing Big Brother and his clones?' asked Buck.

'I don't think revolution will help our situation,' I said. 'The Australian economy could switch from factory goods to personal services while maintaining employment. Houses, cars and consumer goods can be taxed, recognising that private ownership is a privilege and makes demands on energy and water supply. Formation of wealth and prosperity seems inextricably linked to energy consumption. More equitable remuneration is necessary.'

'Next reader, please.'

We can conclude that our society demands and uses energy in many ways, as equally as our democracy will allow. Energy is not the same as money, but it is a commodity with markets and has its own economy, now become political economy. In the first chapter, I speculated that peak oil was a balance between burgeoning demand and restrained depletion. It seems possible that human energy use will reach a plateau rather than a peak, whereafter demand will be eked out and sustained not by physical resources but by non-energy interests.

For example, overseas travel could diminish because foreign travel can be simulated and therefore is real. According to philosopher Jean Baudrillard, humans are embracing hyper-reality and simulation.

Sports games are reproducible and made real by unique outcomes. For the same reasons, fiction, musical performances and theatre for the masses will give way to recordings and movies. Authentic experiences may only be available to the bourgeoisie, whose privilege will include foreign travel and cultural events but their energy greed could be rejected. Regression to the mean will dominate in social relations maintaining cohesion in society and preventing energy greed such as jet travel.

Lives of the proletariat that have been dominated by images will transfer to objects acquired in groups, like laundromat use and gaming. The spectacle has commodified energy supply and now simulation is representing energy demand.

For lives to be fairer, more equality and an end to growth is essential.

By exporting coal we can bring electrification to underdeveloped countries, causing an improvement in living standards more equally than by exporting wealth generation industries.

'That's our last lesson on Energy this year. I will continue next year with The Living World and Technology.

'You have received your exam results. Most of you did well.

'I hope you have enjoyed the course and have learned something useful,' I said jovially.

'Yes. It has been good. Thank you, sir,' said Tracey, as she left. 'I'm looking forward to next year's course.'

'It's been a pleasure, Tracey.'

The students put their hands together and there were whistles and whoops.

TASK 42
What trends in energy supply are there?

TASK 42 Answers
Commodification and simulation.

Animal Farm 2 - satirical fiction

Animal Farm 2 continues Animal Farm, George Orwell's 1945 political satire, updating it within a broader context of the Cold War and its aftermath, with superpowers' environmental movements.

The farm is on tropical Caruba, an island controlled by the Social Republic near the Democratic Union, who are in a Cold War with them. Pigs lead an animal rebellion then takeover and ruthlessly exploit the animal workers.

When the farm animals discover coal on the farm and mine it to supply electricity from a power station constructed on the farm, they become embroiled in the superpowers' climate manoeuvring.

When coal mining could be stopped due to global warming fears, the animal workers study climate science and realise that they are victims of world superpower totalitarianism.

They prepare to fight for animal liberation, wanting freedom at any cost.

The satire is humorous with animal characters based on leaders of superpower nations, animal liberation and climate campaigns.

Farm animals investigate philosophies of climate science within a new paradigm.

A satirical fiction novel, sequel to Animal Farm by George Orwell, 1945.

A Novel by: **Martin Knox**

www.amazon.com.au/Animal-Farm-2-Martin-Knox/dp/0648993027

Will the nanny state and the media industry in future restrict all types of individual performance?

Chance struggles on a capitalist employment treadmill that denies him freedom to take risks. He starts a PhD in science and psychology.

His girlfriend Megan is a pole vault champion in training for the 2032 Olympic Games in Brisbane. With his help, she uses philosophies of De Beauvoir, Nietzsche, Heidegger's phenomenology and Mihaly's flow, to take personal responsibility for her training.

Overreach by the nanny state threatens her use of flow. 'Levellers' want all ability levels to succeed in competitions. Philosopher Debord's 'spectacle' has performers' appearances profiting media and investors.

Which will succeed: human turkey-like individualism or bee-like collectivism?

When Chance and Megan catch Covid, they oppose the collectivists' mandatory vaccinations with non-violent civil disobedience. Will they succeed?

This is speculative fiction satirizing athletics, rugby and ballet, exposing dark forces shaping western democratic societies, revealing how individual freedom can be saved.

Can Chance and Megan overcome a future nanny state's overreach by passive resistance?

TURKEYS NOT BEES

Martin Knox

www.amazon.com.au/Turkeys-Not-Bees-Martin-Knox/dp/0648993043

www.ingramcontent.com/pod-product-compliance
Lightning Source LLC
Chambersburg PA
CBHW072154070526
44585CB00015B/1137